Image Beyond the Screen

Series Editor
Sylvie Leleu-Merviel

Image Beyond the Screen

Projection Mapping

Edited by

Daniel Schmitt
Marine Thébault
Ludovic Burczykowski

WILEY

First published 2020 in Great Britain and the United States by ISTE Ltd and John Wiley & Sons, Inc.

Apart from any fair dealing for the purposes of research or private study, or criticism or review, as permitted under the Copyright, Designs and Patents Act 1988, this publication may only be reproduced, stored or transmitted, in any form or by any means, with the prior permission in writing of the publishers, or in the case of reprographic reproduction in accordance with the terms and licenses issued by the CLA. Enquiries concerning reproduction outside these terms should be sent to the publishers at the undermentioned address:

ISTE Ltd
27-37 St George's Road
London SW19 4EU
UK

www.iste.co.uk

John Wiley & Sons, Inc.
111 River Street
Hoboken, NJ 07030
USA

www.wiley.com

Cover image: Copyright of Cité des Électriciens, Bruay-La-Buissière (France) © Rencontres Audiovisuelles – Video Mapping Festival 2018.

© ISTE Ltd 2020
The rights of Daniel Schmitt, Marine Thébault and Ludovic Burczykowski to be identified as the authors of this work have been asserted by them in accordance with the Copyright, Designs and Patents Act 1988.

Library of Congress Control Number: 2019952956

British Library Cataloguing-in-Publication Data
A CIP record for this book is available from the British Library
ISBN 978-1-78630-504-6

Contents

Foreword xiii

Introduction xvii

Part 1. History and Identity. 1

Chapter 1. The Origins of Projection Mapping 3
Ludovic BURCZYKOWSKI

    1.1. Introduction. 3
    1.2. Let's moonwalk! A short crossing through time 4
        1.2.1. The emergence of the expressions "video mapping",
        "projection mapping", "spatial augmented reality" and "spatial
        correspondence" between the beginning of the 21st Century
        and the end of the 20th Century 4
        1.2.2. From 17th Century magic lanterns to ancient camera obscura . . . 5
        1.2.3. The screen as a material considered as a void:
        projection mapping in negative from the 15th Century onwards 6
        1.2.4. How far back in history can we go? 7
    1.3. Immersion in hallucinated worlds. 8
        1.3.1. Some films on the theme of nested or fallacious
        realities in line with the first digital projection mapping installations . . . 8
        1.3.2. Some philosophies of illusion 9
    1.4. Examples of visual devices 10
        1.4.1. Two visual instruments: anamorphoses and X-rays. 11
        1.4.2. Immersive panoramas 11
        1.4.3. Augmented reality and low-tech virtual reality 12
        1.4.4. Some visual sequences spatialized since Antiquity 13
    1.5. The agencies 14

1.5.1. The arts of memory. 14
1.5.2. Feedback, or the chicken and the egg problem. 15
1.5.3. Some practical uses of the magic lantern 16
1.6. A figure of transgression and juxtaposition with a beyond 17
1.6.1. Unconditionality 17
1.6.2. Magic image imagery 18
1.6.3. Anima. 20
1.6.4. See from a distance. 20
1.7. The invention of an "empty box" as an image container 21
1.7.1. Any precursors?. 22
1.7.2. Alberti and the invention of the screen 22
1.7.3. The humanistic context of the disruptive
object-subject disconnect reified in and through the image 23
1.7.4. A hypothetical starting point 24
1.8. Modern inflexions: obsolescence of old visual devices
and tacit challenges to the Albertian model. 25
1.8.1. Obsolescence 25
1.8.2. Challenges 26
1.9. Parastatic scenography 28
1.9.1. For the eyes: the uncomplicated image 28
1.9.2. Living presences and images 29
1.9.3. From the screen to film 30
1.10. From expedition to investigation 32
1.10.1. Resilience. 32
1.10.2. Ongoing investigation 33
1.11. Conclusion 34
1.12. References. 34

Chapter 2. The "Spatialization" of the Gaze with the Projection Mapping Dispositive 37
Justyna Weronika ŁABĄDŹ

2.1. Introduction. 37
2.2. The release of the "cinematographic cocoon" 38
2.3. Changing the projection mapping dispositive. 41
2.4. The spatialization of the gaze or the perception
of the projection mapping spectator 44
2.5. "Attractions set-up" or real content? 48
2.6. References 49

Chapter 3. Projection Mapping: A New Symbolic Form? 51
Martina STELLA

3.1. Introduction. 51

3.1.1. Symbolic form and apparatus	51
3.1.2. Apparatus and projection mapping	53
3.2. A shifting tool	54
3.3. The surface	56
3.3.1. The environment/projection ratio	56
3.3.2. The volume	57
3.3.3. The projection plane: the substrate	59
3.4. The projection	60
3.4.1. The haptic image	60
3.4.2. The point of view or the projector	61
3.5. Conclusion	63
3.6. References	66

Chapter 4. Points of View: Origins, History and Limits of Projection Mapping 69
Ludovic BURCZYKOWSKI and Marine THÉBAULT

4.1. The origins of a movement towards alternative forms according to Romain Tardy	69
4.1.1. Origins and VJing	69
4.1.2. Transformation and continuity	70
4.1.3. Projection mapping and the screen	71
4.1.4. Projection mapping of yesterday, today and tomorrow	72
4.2. A short history of projection mapping according to Dominique Moulon	73
4.2.1. Projection mapping in the history of light	73
4.2.2. The invention of the video projector	74
4.2.3. The feeling of immersion with different applications of projection mapping	75
4.2.4. The role of ICTs today and tomorrow	77
4.3. Projection mapping and its limits according to Christiane Paul	78
4.3.1. The New Aesthetic	78
4.3.2. Projection mapping as a technology	79
4.3.3. Projection mapping as an experience connecting the physical and the virtual	80
4.3.4. Projection mapping and museums or art institutions	81

Part 2. Texts and Techniques 83

Chapter 5. Listening to Creators in Residence 85
Marine THÉBAULT and Daniel SCHMITT

5.1. Creators, a residence and a festival	85
5.2. Capturing the genesis of a work	86

5.3. REMIND: a method to capture the dynamics
of the situated creative experience 87
5.4. Space, tool and solitude. 88
    5.4.1. The instrumental space. 89
    5.4.2. The dynamics of the emotional states of the creators *in situ*. 97
    5.4.3. Work, emotions and troubles 99
5.5. New residence arrangements 100
    5.5.1. Limitations and contributions of this type of survey 100
    5.5.2. Towards a design of space and experience 100
    5.5.3. The creator profession 101
5.6. Prospects for the future 102
5.7. Increased attention to the place of creators in digital arts 103
5.8. Acknowledgements 104
5.9. References 104

Chapter 6. Projection Mapping and Automatic Calibration: Beyond a Technique 107
Sofia KOURKOULAKOU

6.1. Introduction. 107
6.2. Towards a new projection dynamic. 107
6.3. Automatic calibration 108
6.4. Automatic geometric calibration 109
    6.4.1. Procams methods 109
    6.4.2. Zhang method (Zhang 1998, 1999) 109
6.5. Projector calibration using one or more pre-calibrated cameras 109
    6.5.1. Fringe Pattern/Structured Light DMD
    (Digital Micromirror Device). 110
6.6. Automatic calibration applied 111
6.7. Automatic calibration in France. 112
6.8. Conclusion 112
6.9. References 113

Chapter 7. Projection Mapping Gaming 115
Julian ALVAREZ

7.1. Introduction. 115
7.2. Specifying the scope of the projection mapping game 118
7.3. The indoor projection mapping game 119
7.4. The outdoor projection mapping game. 123
7.5. Conclusion 125
7.6. References 126

Chapter 8. Projection Mapping and Photogrammetry: Interest, Contribution, Current Limitations and Future Perspectives 127
Nicolas LISSARRAGUE

    8.1. Introduction. 127
    8.2. State of the art 127
    8.3. Photogrammetry for projection mapping 129
    8.4. Contribution: an automated imaging device
    for object photogrammetry 130
    8.5. Current limitations and future prospects. 137
    8.6. References 139

Chapter 9. Points of View: Sound, Projection and Interaction 141
Jérémy OURY, Ludovic BURCZYKOWSKI and Marine THÉBAULT

    9.1. Sound creation projection mapping, a real composition of sound 141
        9.1.1. Introduction 141
        9.1.2. The place of sound 142
        9.1.3. Analysis of works of art 146
        9.1.4. Conclusion. 149
    9.2. Projectionist: a profession according to Pascal Leroy 150
        9.2.1. History 150
        9.2.2. Identity and tastes. 151
        9.2.3. Art and technology 151
        9.2.4. Limitations. 152
        9.2.5. Projection mapping and cinema 152
    9.3. Interactive projection mapping by Anne-Laure George-Molland 153
        9.3.1. Enter interactivity to make it exist 153
        9.3.2. Small interactivity and projection mapping 155
        9.3.3. The future of interactivity in projection mapping 156
    9.4. References 157

Part 3. Production and Dissemination 159

Chapter 10. The Factory of the Future, Augmented Reality and Projection Mapping 161
Pascal LEVEL

    10.1. Introduction 161
    10.2. The factory of the future 161
        10.2.1. The process. 161
        10.2.2. The technological challenges of the plant of the future 163
        10.2.3. A digital and connected factory. 164
    10.3. Augmented reality 165

10.3.1. Simple definition 165
10.3.2. Some chronological references for augmented reality 166
10.4. Factory of the future and augmented reality 169
10.5. Augmented reality and projection mapping 170
10.6. Future plant and projection mapping 171
    10.6.1. Some preliminary considerations. 171
    10.6.2. Some examples of projection mapping in manufacturing 172
10.7. Conclusion 175

Chapter 11. Heritage Mediation through Projection Mapping 177
Alexandra GEORGESCU PAQUIN

11.1. Introduction. 177
11.2. The symbolic value of heritage 179
11.3. Projection mapping as a means of cultural heritage mediation 180
    11.3.1. Transcending mediation 181
    11.3.2. Combined mediation 186
    11.3.3. Self-reflective mediation. 189
11.4. Conclusion: monumentalize the monumental 194
11.5. References. 196

Chapter 12. Projection Mapping: A Mediation Tool for Heritage Resilience? 199
Hafida BOULEKBACHE and Douniazed CHIBANE

12.1. Introduction. 199
12.2. Architecture, a heritage trace and an art to be preserved. 200
12.3. The architectural heritage between preservation
and mediation issues 203
12.4. Meeting between architectural heritage and projection mapping. . . . 203
12.5. Classification of architectural projection mapping 205
    12.5.1. Communication issue. 205
    12.5.2. Information issue 208
12.6. Meeting between architecture and projection mapping 209
12.7. Conclusion 210
12.8. References. 211

Chapter 13. Architectural Projection Mapping Contests: An Opportunity for Experimentation and Discovery 213
Jérémy OURY

13.1. Introduction. 213
13.2. Different projection mapping projection contexts 214
    13.2.1. Limitation of projection mapping orders 214
    13.2.2. Contests, platforms of creative freedom. 215

13.3. Interests and functioning of the contests 216
   13.3.1. The organizers' point of view 216
   13.3.2. Functioning of the contests 217
13.4. Analysis of the 2018 season 220
   13.4.1. Perspective of the artists 220
   13.4.2. Results of the 2018 contests 223
13.5. Conclusion 226

**Chapter 14. Points of View: Supporting and
Highlighting Projection Mapping** 229
Marine THÉBAULT and Ludovic BURCZYKOWSKI

14.1. Video Mapping European Center according to Antoine Manier 229
14.2. Lighting design and sustainable projection
mapping installations according to Alain Grisval 231
   14.2.1. Lighting designer 231
   14.2.2. Durable devices 232
   14.2.3. Economy 232
   14.2.4. Legal aspect 233
   14.2.5. Identity and taste 233
   14.2.6. Interaction for all audiences 234

List of Authors 235

Index 237

Foreword

It is a well-worn idea that this century is one of an abundance of images. It is another to say that we live in an era of screens: for a long time, we had two in our lives – cinema and television. We then went from three, with the computer screen that opens up access to the immense resources of the Web and the Internet, to four with mobile tools – smartphones. Some are pursuing inflation by adding five (the intermediate terminal that is the tablet), six (the games console) and seven (the immersive helmet). Little by little, screens are saturating all the dimensions of our private but also public spaces, as we can see with the recent replacement of traditional advertising billboards by screens in our developed cities – this, moreover, without any consideration for the ecological impact that this mutation implies. The screen is everywhere.

This book opens a much less widespread space: the space of the *image beyond the screen*. From monumental projections on façades to more intimate projections on sets or objects. Is it worth stopping by? Isn't this just a simple change of media? An additional avatar in the diversification of screens?

Juxtaposing chapters by researchers and testimonies by artists and practitioners experienced in this new form of projection, the texts gathered in this first book on projection mapping show in various ways that, on the contrary, a completely new form of expression, or even an art, is being exhibited.

On the one hand, several chapters remind us of this: Leon Battista Alberti, in *De Pictura* in 1435, presents the painting as an "open window". Inside the frame, the "white canvas" is an immaculate void replaced by an external scene, similar to the opening of a window that allows the inhabitant to see the outside of his residence while remaining inside. The screen takes up this characteristic of painting: in a similar way, it is a window open onto a landscape and/or a scene to be inserted, totally independent from the place where it is located. The situation is quite different with projection mapping, which brings out the image on the set, and where the specific geometry of the projection medium reappears: it is essentially a question of "dealing with" it, and not of going through it to replace it with something else that has no connection and that ignores it.

On the other hand, the screen is also defined by immobility. It cuts a window of escape into the space and time of reality. In cinema or video, the reported scene that fits into the window is dynamic and has movement, but the window itself is motionless. Moreover, the inscription of the reported image takes no account of the environment, as shown by Rem Koolhaas' caricatural aphorism about *the context*, which Ludovic Burczykowski recalls in Chapter 1. On the contrary, projection mapping develops a new mediation mechanism, that is "an action involving a transformation of the situation or the communicational mechanism, and not a simple interaction between elements already constituted, and even less a circulation from one element from one pole to another" (Davallon 2004, p. 43), as Alexandra Georgescu Paquin states in Chapter 11.

Thus, projection mapping is an emergence in the sense of Morin (Morin 1977; Juignet 2015): neither the medium, nor the projection, but an in-between. Or rather a third-party composition that could not survive without the medium or projection. A type of link, of junction from which the unexpected springs forth. This is what Martina Stella refers to as a subject-matter in Chapter 3. In this form of "narrative composition", the reciprocal action of one on top of the other creates movement on a monumental medium. This is particularly the case in these ever-impressive sequences where the building explodes or collapses, or when a window appears that opens into a well-known facade that is known to be blind. In another style, this is also the case when the image animates a white marble statue, making it cry or laugh, changing its expression at will, following the viewer's gaze or winking at him. And we start dreaming of marble statues that might start dancing!

Last but not least, the way in which monumental projection mapping is broadcast brings back to life a practice that is in retreat, that of strolling and street performance. During the second projection mapping festival in Lille in March 2019, more than 100,000 people walked for an entire evening between the spots spread throughout the city[1]. And it is no coincidence that this form of projection is so successful in this region of Hauts-de-France, which has always loved the big popular parades and street carnivals. The device imposes the social sharing of a collective viewing in the public space, far from the intimate, internalized and silent contemplation specific to cinema, or the superficial and distracted viewing that can be practiced alone or in groups in front of television or on mobile devices.

Finally, projection mapping embodies in a new way the idea of a *Gesamtkunstwerk* dear to Richard Wagner. Indeed, sound is part of the projection mapping work, just like the image. It is therefore a form that links the object and/or the architectural environment with the audio-visual media. But experiments have already given way to gesture in the construction, such as this work which translates into images the gestures of the conductor conducting the musical work performed live. Taste and touch may follow in the dynamics of innovation.

In the end, we can only agree with Antoine Manier when he states that projection mapping works are in their infancy and will be structured. A grammar of projection mapping will be gradually developed. To master it, it will be essential to simultaneously understand how projection mapping makes sense, and what the viewer develops from the perceptual requests addressed to him (Leleu-Merviel 2018). Gradually, we will be able to move towards modelling in support of the design of works, as we were able to do some 20 years ago for hypertexts and hypermedia (Durand *et al.* 1997). Other techniques will emerge, such as photogrammetry (Chapter 8), while other algorithms will refine the correspondence calculations between the 3D surface and the mapped image, and other fields of implementation will be strengthened, such as the factory of the future (Chapter 10). Other concepts

---

1 There was no certified count, but estimates, based on objective criteria (for example, the number of people entering indoor spaces with limited capacity, or number of connections to the mobile application), cite around 130,000 participants.

will structure the reflection about the image beyond the screen, such as mediatecture[2] (Chapter 11). Undoubtedly, this book is only a foundation which many others will come to complete before we have covered the subject.

Sylvie LELEU-MERVIEL
Director of the DeVisu laboratory
Polytechnic University of Hauts-de-France
Valenciennes

References

Alberti, L.B. (1435). *De Pictura*. Macula, Paris.

Davallon, J. (2004). La médiation: la communication en procès? *MEI*, 19, 37–59.

Durand, A., Huart, J., Leleu-Merviel, S. (1997). Vers un modèle de programme pour la conception de documents. *Hypertextes et Hypermédias*, 1(1), 79–101.

Juignet, P. (2015). Edgar Morin et la complexité. *Philosophie, science et société* [Online]. Available at: https://philosciences.com/philosophie-generale/complexite-syste me-organisation-emergence/17-edgar-morin-complexite.

Kronhagel, C. (2010). *Mediatecture: the Design of Medially Augmented Spaces*. Springer, Vienna.

Leleu-Merviel, S. (2018). *Information Tracking*. ISTE Ltd, London, and Wiley, New York.

Morin, E. (1977). *La méthode 1. La nature de la nature*. Le Seuil, Paris.

---

2 "Mediatecture is the orchestration and temporalization of space, the loading of meaning into space and the creation of a sphere of communication" (Kronhagel 2010, p. 3).

Introduction

It is generally agreed that projection mapping consists of applying light whose geometry corresponds to a more or less complex surface made of heterogeneous materials on which it is projected. Is it really that simple? The practice of projection mapping has developed since the 1990s, particularly in the artistic field, by supporting the dynamics of digital technologies. This practice has now become almost commonplace, while paradoxically, research in the social sciences and humanities pays little attention to it. Certainly, projection mapping could be understood as a simple hybridization of cinema, animation and scenography, but that would miss the essential questions that this practice stimulates.

First of all, the very concept of projection mapping needs to be clearly defined: is it a tool, a device, a technique, a medium, a discipline, a practice, a trend or a movement? Can an image do without a connection to any surface? Projection mapping is synonymous with *spatial augmented reality*, *video mapping* and *spatial correspondence*. While the words video, projection, reality, augmented, spatial and correspondence have found their French equivalents, the word *mapping* presents itself as an obstacle to translation. Few say that they do "video cartography" or "projected cartography". The term "mapping" refers to dressing, coating, texturing, covering, transposing and tuning. A rigorous translation of projection mapping as a medium in our context could then be "projection of correspondences" or "video-projected correspondences".

Introduction written by Daniel SCHMITT, Marine THÉBAULT and Ludovic BURCZYKOWSKI.

In this book, many people use the expression *projection mapping* as we have retained it. Creators, broadcasters, sponsors, service providers, producers, audiences, enthusiasts, technophiles, critics from various disciplinary backgrounds define it and recognize themselves in its lexical field.

The Video Mapping European Center project, supported by Rencontres Audiovisuelles and the DeVisu laboratory of the Polytechnic University of Hauts-de-France, offers and organizes training courses, artist retreats, screenings, a festival and a conference dedicated to projection mapping: *Image Beyond the Screen International Conference* (IBSIC). This project has informed this book, which aims to identify the different conceptions and practices of projection mapping in order to give it a status, identity, issues and perspectives. For us, it is a question of setting a milestone in favor of its theoretical existence as a medium and discipline in order to stimulate reflection, propose a new perspective on the scriptures, and enrich future practices. The research and points of view proposed in this book contribute to the construction of a shared theoretical object that was previously lacking. It is divided into three parts:

1) history and identity;

2) writing and techniques;

3) production and dissemination.

As far back as we can go in the history of light projection devices in space, it remains difficult to define precisely the genesis of projection mapping (Burczykowski). A practice that remains close to cinema and animation, but which is not totally in line with it. The screen, fixed, mobile, inert, living, pre-existing to the project or created specifically, is at the heart of the matter (Alvarez; Tardy). Video mapping is available in XXL format as well as in miniature, transportable and intimate format, perhaps promoting interactivity (George-Molland). It is far from being systematically narrative and offers transmedia and sustainable opportunities (Grisval). It often has audio to develop the contours of the world that the work seeks to create (Oury). The invention and subsequent democratization of the video projector and digital tools have paved the way for new forms of expression (Moulon). Artists have sometimes made the city their studio, their relationship to the workplace being essential in the creative process, which is far from being a long quiet river (Thébault and Schmitt). Other technological opportunities such as the automation of the geometric calibration of projected images (Kourkoulakou) and photogrammetry (Lissarrague) allow time saving during content creation by way of an accelerated evolution of methods and uses. A society can be defined through the evolution of its appliances and devices, but above all the appliances build the sensitive potential of a society (Stella).

Projection mapping is being developed in multiple ways. It serves the industry of the future by guiding complex and delicate assemblies of machined parts (Level). Based on a low-intrusive creative action, it transforms a heritage or updates a memory trace (Georgescu Paquin); it highlights content that enhances the medium (George-Molland). In architecture, projection mapping is becoming a learning device that promotes reflexivity and memorization (Boulekbache and Chibane). It offers game experiences (Alvarez) and although projection mapping writing is in its infancy, it will probably structure and develop (Manier). Experiments, whether aesthetic, technical or mediation, will be expanded. This audio-visual form is born with the evolution of needs, disrupting and broadening our perceptions (Labadz).

Image technologies influence the way we represent ourselves. They extend our body and allow us to visualize it (Paul). Projection mapping also offers a special experience; it is a source of amazement and wonder (Leroy). Most often, the illusion created by the projection does not require any equipment on board the spectators. The festival is now one of its main forms of expression, particularly in South America (Oury). For as much as and beyond the major events, electronic, technological and digital art is now spreading to institutions and galleries (Moulon). This is not without influence on their presentations and documentation of the collections and their conservation (Moulon; Paul).

We hope that these points of view will contribute to giving projection mapping a theoretical consistency, and thus accompanying and enriching the future of what could well become a new art, a new discipline, and a new field of research.

PART 1

History and Identity

1

The Origins of Projection Mapping

1.1. Introduction

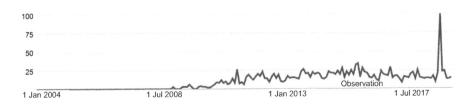

Figure 1.1. *Evolution of the queries of the keyword "video mapping" in French on Google from 2004 to 2019 (source: Google Trends, 2019). For a color version of the figures in this book see, www.iste.co.uk/schmitt/image.zip*

When did projection mapping start? More than 10 years separate this text from the emergence of the key word "video mapping" on Google, but this criterion alone is not enough to capture its origin and requires a more distant archaeology. The materials that make up this chapter come from a collection of significant historical elements during a series of expeditions in the reflector that is the archived history of ideas and works of art or technology up to the 20th Century. Presented with a desire to go beyond the passive collection, these facts are ordered in such a way that one can understand how projection mapping can be understood as an instrument of vision comparable by similarities and differences to many other devices that

---

Chapter written by Ludovic BURCZYKOWSKI.

preceded it, but also as an original way of seeing that has been maintained over time.

1.2. Let's moonwalk! A short crossing through time

1.2.1. *The emergence of the expressions "video mapping", "projection mapping", "spatial augmented reality" and "spatial correspondence" between the beginning of the 21st Century and the end of the 20th Century*

By literally writing "video mapping" on the Google Trends trend analysis tool, we can see that queries for this keyword increase significantly from a point of origin located in 2008. This information highlights the recognition of the unity of a concept based on Web searches for it and would make it possible to clearly date the birth certificate of projection mapping. Google and the use of this word alone, however, do not reflect what precedes them: before 2008, this expression was not used or very rarely used, whether written as one or two words, with or without an accent, and with or without the word "projection" instead of "video". This initial year would therefore be sufficient if it was not very clear that the technical device and its spatial writing logic were already well identified in the 1990s by technologists or artists. The often cited "spatial correspondence" was first mentioned by Michael Naimark in 1984 in Los Angeles, while "spatial augmented reality" was first mentioned in 1998 in North Carolina by Ramesh Raskar.

Between these two publications, Paul Milgram proposed in 1994 in Toronto a taxonomy of "mixed realities", and the first patents identified for digital video projections that match volume surfaces were filed for Disney Inc. in 1991 and General Electric Co. in 1994. Between these two patents, artist Tony Oursler exhibited *The Watching* in 1992. Later, in 1999, John Underkoffler, the designer of *Minority Report*'s famous interface, invented *I/O bulb* and *Luminous room* when the facade of Amiens Cathedral was painted. In 2004, choreographer Klaus Obermaier used infrared to divert the real-time video from the silhouette of a performer in the foreground of the stage and projected a video on his body that differed from the background. Johnny Chung Lee published his doctoral thesis on automated projector calibration the same year. The following year in 2005, Olivier Bimber planned to project onto a classical-age painting.

1.2.2. *From 17th Century magic lanterns to ancient camera obscura*

If we stick to the technical device and spatial writing logic, many artists, painters, visual artists, illustrators or directors have also used, since the beginning of the 20th Century, the unconventional image projected in relation to the reference that is cinema. They played with the image carriers and environments of their works in a way that projection mapping still offers today. Before them, in the 19th Century, it was travelling projectionists like Maximillian Skladanowsky who projected natural disasters, sometimes improvising projection spaces for storms, fires or earthquakes. These images are reminiscent of buildings that collapse with digital pixels. Savoyard lanternists brave the mountain slopes accompanied by marmots, monkeys, drums, or barrel organs like the contemporary off-road projection mapping projectionists do outside rooms.

At the end of the 18th Century, phantasmagorical spectacles such as those performed by Giuseppe Balsamo called the Count of Cagliostro, Johann Georg Schröpfer, Paul Philipsthal or Étienne-Gaspard Robert called Robertson used illusory projections of supernatural beings. They would double the projections of thunderous noises or the smell of burnt feathers. Between the 15th and 16th centuries, a magic lantern was designed by Leonardo da Vinci, but they were not made by Christiaan Huygens until 1659. Then they were carried around in 1664 by Thomas Walgenstein, at the time when anamorphoses were theorized. The lantern is widely disclosed by Johann Christoph Sturm and the Jesuit Athanasius Kircher. It was Milliet Dechales, in 1674, who produced a series of projected images approaching animation and, in 1698, Johann Christoph Weigel developed an overlay of images in the same way as a compositing of layers. This was shortly before the lantern was mounted on wheels and several were used to compose the same image, which today would be called stacking or shifting the image. Johannes Zahn, a Jesuit, also shows animated projections: he projects worms in a jar, the movements of a weather vane or those of a clock. Van Musschenbroek continued Zahn's work around 1730 by inventing animated plates that made it possible, for example, to turn the wings of a mill. As early as 1608, Cornelis Drebbel said he could change the appearance of his clothes and make giants appear.

In 1650, another Jesuit, Gabriel Magalhaens, reported descriptions of the use of oriental lanterns that he said were quite convincing. Chiang Khuei and

Fang Chheng, in the 13th Century, made animated projections on smoke. Sun Kuang-hsien, in 930, or Shao Ong the magician, as early as 121 BC, did comparable tricks. In the past, they would turn flying dragons. In the 16th Century, in the West, Giambattista della Porta described in his *Magie naturelle* how to reveal volumetric light in chalk powder. He provides a remarkable drawing of it. Girolamo Cardano and Benvenuto Cellini projected images for shows. In 1420, Johannes de Fontana produced a drawing of what he called the lantern of fear in the *Bellicorum instrumentorum liber*. Jokes and mysteries are played in the Middle Ages with these thaumaturgical lanterns. Arnaud de Villeneuve, from 1290, would have used it. The *camera obscura* was known to Aristotle and Mozi in the 4th Century BC, and even to Apollonius of Tyana, the miracle worker, in the 1st Century.

Contemporary archaeologist Matt Gatton argues that the *camera obscura* may have been used as an archaic projector to animate the faces of statues during the Eleusian feasts in ancient Greece – no doubt long before the use of 16 mm projector film from 1969 onwards to project on the busts sculpted inside Disneyland's *Haunted Mansion* and on their facades and other crystal balls. Contemporary authors Jean-Jacques Lefrère and Bertrand David have even put forward the hypothesis, with supporting arguments, that figurative shadows were projected using tallow candles as early as the Palaeolithic era.

1.2.3. The screen as a material considered as a void: projection mapping in negative from the 15th Century onwards

As far as we can go in the history of light or shadow projection devices, pinpointing a beginning seems complex. Would it be necessary to have a kind of visual antinomy of what projection mapping is in order to be able to record its beginning by way of a contrast? One can find an antinomy of this type in an asserted disregard for the medium receiving the image, since it is the interplay with the latter that gives meaning here to the words *mapping*, *correspondence* or *spatial*. One of the most important key points of such disregard is undoubtedly the explicit verbalization of Leon Battista Alberti written in 1435 in *De Pictura*. He presents the painting board as an "open window". Mapping on the void that is the open interior of a window is a misdirection: no more composition according to the medium is possible as soon as it is expressed as non-existent.

If we agree to consider the artistic movement that began with the transition to the 20th Century of so-called *modern* art – or at least one of the most important parts of this movement – as a challenge to this Albertian representation as a visual screen form that has become hegemonic, the hypothesis is that projection mapping would only be one manifestation, among other types, of what happened as a result of this paradigm shift. Extension of the painting and access to new dimensions sought by painters, removal of the base in sculpture, total or synesthetic art, sublimation or enhancement of the material, background scenery becoming an actor in scenographies, etc., are all fine art studies whose origins can be traced to projection mapping.

1.2.4. *How far back in history can we go?*

Thus, the first image not to consider its medium as an "open window" is the one that took advantage of the formal specific features of its surface, considering the material nature or volume of the latter. We then find ourselves in the astonishing situation of having to bring together the origin of projection mapping with that of the images themselves at the current state of knowledge of the history of human creations: wall representations of the Paleolithic. Some of them undoubtedly play with the volumes of the cave walls, which is a form of *mapping* in the sense used in projection mapping: a set of correspondence between an image and the specific features of a heterogeneous medium. Some of these images could be animated with the moving flame of a fire, in a sense close to what can be understood as the feeling of movement caused by the video. In addition, some of them took place where acoustic games were played. Would a study on projection mapping help to understand this mystery? It would have been easy to conclude that projection mapping was anthropologically inevitable since it was so latent in humanity that it became confused with it. Everyone dresses, every culture transforms or disguises their bodies, makes them up or masks them. What is more than mapping if not disguising a surface by pretending to transform it by adding light? Exposing it, perhaps.

So, what alternatives occurred when the Albertian representation was set up? Was there any kind of projection mapping before it was designated? Would we have always done it? If so, the words *video* or *projection mapping*, *spatial correspondence*, *spatial augmented reality*, would use new terms to describe something old that does not have this name, and this would legitimize an archaeological expedition. But if we insist on going back in time, we notice that projection mapping becomes a concept whose strength, like any good concept, is to find many dynamic echoes in various contexts. The difficulty of such an undertaking, then, is to know where to stop the investigations knowing that we will always find if we want a Greek, a Chinese, an Indian, an African tale or a disappeared civilization which will have made some initial progress. Or, at the very least, we could build a demonstration of it! The risk of seeing the new in the old is equal to the ambition when it comes to making the new from the old. Let's extrapolate, then: hardly earlier than the first shapes drawn in the caves, we would make the *White hole, White fountain* or *Big Bang*, the first "projectors" of light, cosmic! Are we living in an illusory *Matrix* projection? To pose this question, we must take a look at some media and philosophical doctrines.

1.3. Immersion in hallucinated worlds

It was shown in September 2017 that the resources of a conventional computer do not allow the known universe to be simulated. A few months earlier, an experiment was attempted with the Swiss supercomputer *Piz Daint*, which provides an image of the universe composed of 15 billion galaxies generated from two trillion particles in 80 hours of computing. This representation of a closed world is nevertheless in contrast to the infinitization of the universe that marks the 16th Century described by Alexandre Koyré, but the idea of the anthropocene today also brings us back into a relatively closed system. Sustainable development problems linked to climate change or access to natural resources thus tend to show us our contemporary artificial environment as an ephemeral illusion in view of the changes deemed necessary for the future. To get an idea of this, it is sufficient to look around you, identifying everything that has required oil, comparing predictive studies on its extraction and consumption with the materials that you imagine you could produce in its absence. This appreciation of a hallucinated world refers to an idea that is not new and crosses history along various multicultural traditions: living in a coma, a dream, as a living being, an egg or a seed that is virtually a tree, a computer simulation, being just a brain covered with electro-stimulators, being possessed, etc., are as many cases of the abyss of illusory worlds where parallel realities of a geometry that is not very Euclidean can overlap or intertwine with each other.

1.3.1. *Some films on the theme of nested or fallacious realities in line with the first digital projection mapping installations*

The novels *Simulacron 3* by Daniel Francis Galouye and *Simulacres* by Philip K. Dick, both published in 1964, are said to have founded the sci-fi theme of the world viewed as a computer simulation. The first one became *Le Monde sur le fil* (1973) and then *Passé virtuel* (1999). The second author would give the essence of *Total Recall*'s schizophrenic scenario (1990). Although Adolfo Bioy Casares' novel *Invention of Morel* published in 1940 was earlier than them, it will not be part of the opuses that adapt this theme on the big screen. In the 1990s, there were *The Lawnmower Man* (1992), *Ghost in the Shell* (1995), *Jumanji* (1995), *Serial Experiments Lain* (1998), *Dark City* (1998), *Truman Show* (1998), the *Matrix* trilogy (1999) based on the novel *Neuromancien* (1984), *ExistenZ* (1999). It was also in 1999 that digital projectors were installed in cinemas, with the digital release of *Episode I* of *Star Wars* as a benchmark. Others followed, including *Avalon* (2001) and *Vanilla Sky* (2001). The latter is a remake of *Open Your Eyes* (1997), a Hispanic film that is itself part of the 17th Century Spanish tradition of playing with the rationality of perceived realities. The most important representatives of this tradition are the novel *Don Quixote* (Cervantes 1605), the play *Life is a Dream* (Calderon 1635) and the painting of *Ménines* (Velasquez 1656).

1.3.2. Some philosophies of illusion

In the East, in the Vedic texts, the *Brahmin* underlies the cosmos as a presence in all things and presents itself as the only reality whose manifestation named *Maia* makes us think of a world that we accept as real. But it remains an illusion to be overcome in order to see the transcendent realities. In Buddhism, which is most probably inspired by it, the ego projects an illusory reality onto a set of primordial laws called *dharma*. It is noteworthy that some Japanese Buddhist temples equipped themselves with projection mapping systems no later than by 2015. Chouang Tzu, in a Taoist fable, wonders if he is a man who has dreamed that he is a butterfly or if, rather, he is not this butterfly dreaming that he is Chouang Tzu. On the western side, the ancient Greeks put forward similar ideas. For the pre-Socratic Heraclitus, around the 5th Century BC, "nature likes to hide itself". The "kora" described in the *Timaeus* by Plato a century later is presented as a dream, both a print and a matrix of a *genesis* perceived also under a veil. Of course, his cave is even more famous, and Plato defended the Pythagorean ideal of an invisible world that could be described mathematically. Galileo summarized this in the 17th Century by saying "Nature is a book written in mathematical language". Shortly after Galileo, it was John Locke and Robert Boyle who imagined that they could bring

Adam and Eve's conditions in Eden back to heaven, thanks to exhaustive properties. Adam and Eve saw things as they *really are*.

The theory of emission, according to which visual perception occurs through light rays emitted by the eyes, has been debated since the ancient Greeks. For Lucretia, in the *De Natura rerum*, written in the 1st Century BC, there is a wandering sham crowd and the vision of the mind coincides with that of the eyes. The fable of the painter Parrhesias winning against Zeuxis illustrates this idea well. In the 18th Century, Francis Bacon wanted to inhibit the tendencies of the mind that he named idols and that distorted reality (*Novum Organum* 1620). Descartes, some 20 years later (*Première méditation* 1641), argues that he has sometimes experienced that "meanings were misleading, and it is prudent never to rely entirely on those who once deceived us". Newton then demonstrates in *Optics* (1704) that what we see is not physical reality. In the 17th and 18th centuries, Berkley, Hume and Kant were famous for their criticism of the idea that reality exists independently of its human representations, known as "realism". Hume, for example, argues that it does not matter whether impressions are produced by the creative power of the mind: what is important is that "we can draw inferences from the coherence of our perceptions, whether they are true or false, whether they represent nature accurately or whether they are pure illusions of the senses".

1.4. Examples of visual devices

Without having to dwell on other more contemporary thoughts, it is already remarkable that the philosophical traditions that characterize the illusory or incomplete state of affairs make it a subject that seems inexhaustible. If we see both too much and too little, if there are things behind things and we perceive only part of them, if we imagine artificial additions, etc., it would be appropriate to inhibit or subject to analysis what potentially deprives us of a proper appreciation. Jean Cocteau thus repeated Tchouang Tseu in his 1950 *Orpheus*: "Some say that we are just his dream", before adding "his bad dream". The idea of a Promethean transgression or *hubris* presents itself as a result, since the challenge of finding some ways to differentiate true and false is symmetrically launched. Should we show ourselves our own ignorance of a fake, hidden or perceived incomplete nature to which we add our own falsehoods? In this sense, Umberto Eco advanced the idea that a process in man leads him to seek iconic processes that increasingly deceive his senses. The following few processes raise this question: would projection mapping answer the need to make reality habitable? With Aristotle, art in some cases completes what nature does not have the power to accomplish. According to Kant, only Art that is evaluated without criteria of utility makes it possible to exercise a free will, and thus to be free. "Augmented reality" would be the fruit of those ideas that have obviously crossed time and space: both an expression of perceptual limits and an apparatus of transgressive vision that tries to approach a beyond an imperfect illusion because it is incomplete and overdetermined.

1.4.1. *Two visual instruments: anamorphoses and X-rays*

The 17th Century Dutch man, living a golden age of painting at a time when we are perfecting the lenses of telescopes, the automation of calculation, the theory of anamorphoses and the diffusion of the magic lantern, does not hesitate to recall the dichotomy of the limits of our perceptions on which a set of devices emerge that aim to overcome them while being conditioned by them. The errors of our vision became a subject of representation for the Dutch of that time and it is in this prism that the theories of anamorphoses which were produced then can be understood. After about 1630, the complexity of the perspective play of the new optics, catoptric and perspective, devalued the human vision now perceived as a simple way of seeing the world among so many others. Jacob Leupold's "machine for anamorphosis" (1713) is an apparatus that shows this devaluation and perhaps even aims to prevent it. A probable high point came in 1895 when Wilhelm Röntgen accidentally invented X-rays. The range of colors whose painting had hitherto celebrated the variety is in fact only a window into the rays that escape our senses. This further mocked the tradition of representation that photography too had already dulled considerably. The mocked painter "can't see further than the tip of his brush".

1.4.2. *Immersive panoramas*

Cinema is this original and radical form of visual immersion that the themes of the nested realities mentioned above translate as a self-reference from its own operating mode. In favor of a certain captivity in the cinema, the cinema does not allow *a priori* any floating attention or bodily mobility. This "immersion" is supported by renowned videographers: Bill Viola when he argues that to see a video image "you have to get wet" or Nam June Paik for whom it is "weightlessness". In both cases, we would no longer have our feet on the ground with the video image. For the theorist of computer graphics, Philippe Quéau, the reason for the enthusiasm for virtual images is also "immersion in the image".

The giant panoramas of the 18th Century precede this radical aspect of an exclusive image. They artificially stage and provoke the sensation of 360° immersion. Around 1779, the engineer of the Royal Corps of Bridges and Causes, Louis Le Masson, had a "great and new" idea "to show Rome as a whole". The device designed is a large circular painting where the viewer's eye finds the most complete pictorial illusion. It was patented in June 1787 by the Irish painter Robert Barker, who named it *La Nature à Coup d'Œil* in French. It is remarkable that the mausoleum of the philosopher Hume, mentioned above as going against the doctrines of "realism", is painted in this first panorama. The year before, the idea of *Panopticon*, Jeremy Bentham's "inspection house" was written, which conceived a circular prison project whose characteristics gave prisoners "the feeling of an invisible omnipresence". Jules Damoizeau (1890) is credited with the first 360° panoramic shot. In another field, Claude Monet began the *Water Lilies* project in 1918. He painted a whole that formed a surface of about 200 m^2 in eight compositions of the same height suspended in a circle.

1.4.3. *Augmented reality and low-tech virtual reality*

These panoramas place their observers in a portion of an imported or imaginary landscape. They are large in size and contrast with smaller portable objects such as Martin Engelbrecht's "perspective boxes" from 1730 or the trompe-l'oeil painter Samuel van Hoogstraten from 1656. In a way, they anticipate today's virtual reality (VR) headsets, as well as Charles Wheatstone's two "stereoscopes" produced in 1838. Brunelleschi already proposed with his "experience" in 1415 to look, through a hole drilled inside a painting, at the reflection of this painting which made it merge with its unframed frame. In the 18th Century, *Claud's Glass* turned its back on what was observed in a tinted mirror to give it the calibrated shape of the chromatic canon of landscape representation of the time. These mirrors prefigure as glasses described by L. Frank Baum in The *Wizard of Oz* (1901) or ocular lenses described by Isaac Asimov in the *Foundation* series which is now called "augmented reality".

1.4.4. *Some visual sequences spatialized since Antiquity*

The history of the first sequences of images placed in space dates back to the images drawn or painted on the walls of caves in the Paleolithic period. Some of them evoke a movement in a succession of simple strokes or poses. The Knowth grave constructed around 5000 BC in Ireland, has symbolic representations engraved on the peripheral stones around its perimeter and may have given rise to circular processions punctuated by visual sequences. A cup found in Iran represents a goat in five different positions around its circumference and dates back to the same period. In the second millennium, in Egyptian-Mesopotamian cultures, frescoes and bas-reliefs by wrestlers were composed of successive poses. In the necropolis of Deir-el-Medineh in Thebes around 1250 BC, the peaceful life in the afterlife is recounted in the style of the characters seen in profile. In the life of the 6th Century BC, Persepolis also had regular sequences of images engraved in bas-reliefs.

Greco-Roman antiquity offers many examples of graphic narrative on various media, whether objects of movable art or real estate architectural art: vases, temple pediments, murals, mosaics. The *Vase François*, which dates back to around 500 BC, is painted with mini-stories that largely refer to mythology. The Trajan column, almost 40 meters high and erected around 110 is also remarkable: nearly 200 meters of bas-reliefs wind around it to tell the story of the conquest of present-day Romania. The Achilles shield described in the Iliad also prefigures for some the discs of 19th Century optical toys, even if it probably never existed.

Pre-Columbian art provides other remarkable bas-reliefs in the Mayan, Aztec or Olmec cultures. Mochica ceramics between the 1st Century BC and the 8th Century break down the rites of this society into graphic sequences. Around the 6th Century BC, after the death of Gautama Buddha, the "stupas", a kind of mound containing his remains, appeared in India. They are first decorated with motifs that meet the need to impart the Buddha's teaching and reveal a true graphic narrative system. This tradition of Asian narrative relief would be expressed in a spectacular way in Angkor, built in the 9th Century.

Projection mapping is sometimes compared to digital wallpaper or tapestry. The *Bayeux Tapestry* was created in Europe in the 16th Century. A long narrative stretching over nearly 75 meters, it is a true documentary that goes so far as to use flashbacks. The oldest example of tapestry is found in the Siberian steppes and dates back to the 5th Century BC. Microcosm representing a garden where the whole world comes to achieve its symbolic perfection according to Michel Foucault, the tapestries of nomadic peoples are mobile and we walk through them in both mind and body. This double practice, the nomadism of its authors and the microcosm represented there, are probably at the origin of the idea of the flying carpet. Far too expensive to be placed on the ground, they are put on objects or walls.

1.5. The agencies

The few visual devices listed above translate what the ancient philosophical traditions have to say about the illusory nature of our perceptions into images. They also inform us that these same devices can be used for a variety of purposes. How do these devices work? What can they produce and what roles do they play?

1.5.1. *The arts of memory*

A structural link seems to combine memory and spatialized images. György Buzsáki and Edvard Moser, in January 2013, in the journal *Nature Neuroscience*, presented the hypothesis of an evolutionary continuity between our cognitive processes for orientation in space and the mechanisms underlying declarative memory. The visual forms of immersion seen above thus evoke another one of a functional and utilitarian nature. A history of images in relation to spaces cannot forget to mention the *arts of memory* that in its palaces or gardens deliberately distribute images in space in a composed way. They were first strictly imaginary before taking shape in concrete achievements. Among the ancient Greeks, the art of memory consists of building a mental architecture that must be populated with characters, objects or scenes that are preferably striking or strange. These images are linked to ideas or words related to what one wants to remember. Subsequently, seeing what happens while walking mentally in this imaginary landscape will bring revive the memory of what is inscribed there.

After Saint Augustine and the 5th Century, there is no longer any explicit reference to this process. It reappeared with scholastics around the 13th Century and cathedrals. We then move from an inner art to a monumental art. In 1534, a simple use of "natural" memory was advocated, although in the 16th and 18th Centuries, Giulio Camillo's *World Theater*, Robert Fludd's *Memory Theater* and Tommaso Campanella's *Sun City* turned the arts of memory into concrete buildings. Its controlled use could give its user supernatural powers. The magician of the time prefigured the modern scientist according to historian Frances Yates. It is Giordano Bruno's task to synthesize this system of place and image with another system, Raymond Lulle's *Ars Magna*. Some people see this as a prefiguration of the animated discs preceding those of the optical toys prior to the cinema. Thirty-four years after Bruno's death, John Bate gave a description of the *Zootrope* in *Mysteries of Nature and Art*. The revolutionary Bruno would not escape the stake in 1600; his system was considered a tool that manipulates phantasms.

1.5.2. Feedback, or the chicken and the egg problem

The fact that visual devices can be used in a utilitarian way makes it possible to implement the idea of these devices having a capacity to act. This will vary according to who owns them, the interest found in shaping perceptions, and undoubtedly the social ground suitable to welcome or not and to transform or not a device or other, or even probably what these devices will generate *by themselves*, if we can say so. It is not clear, for example, who initiated the other from the printing press or the Protestant reform. And if it is clear that Copernicus' heliocentric ideas only caused their upheaval once they were taken up by Galileo's commitment to using telescopes to pass them by, is it not more obvious that it is the telescope itself that pushed Galileo in this direction? The Albertian perspective and its framework screen, which will be discussed again below, raise the same question: does it establish itself by changing its context or did its context establish it?

The visual tool gives a new meaning or sharpens sensitivity, such as the blind man's cane that allows him to imagine a space. With the anthropologist François Sigaut, the tool is also a medium for differentiation with the exteriority and, in this sense, it does not present the tool as an "organic projection" in the manner of Ernst Kapp, whose theory has become so classic that Leroy Gourhan tacitly refers to it. Sigaut, on the other hand, prefers the role of transforming something in us. As Oscar Wilde says, life imitates art much more than art imitates life. Or, to cite Augustin Berque, our technical externalizations come back to us, at the very least, in a symbolic form.

1.5.3. *Some practical uses of the magic lantern*

Developing the concepts of *techno-mimicry* or *technesthesia* in more detail would be welcome if the framework or rather the focus of this text were not the historical contextualization of projection mapping. This text is therefore limited to not developing them and asks us to accept at this stage that projection mapping can change the way we look at objects, or at least to recognize that we can produce new ideas from projection mapping especially for things that are not projection mapping. For example, using a metaphor to illustrate a complex concept or phenomenon. The *camera obscura* reported on the functioning of the eye for Descartes by going beyond the simple metaphor. In a concise summary, it is suggested that if the image of projection mapping is an image transformed and/or composed to match a "reality", this "reality" also comes to match and transform itself according to projection mapping, in what we want to be a virtuous circle. It then finds common origins with visual tools that, from X-rays to spectroscopy or radioscopy through microscopy or film in slow-motion, etc., have played their role in making knowledge accessible or have been diverted to other areas of interest. Sigismund, the central character of Calderón's *Life is a Dream*, reached a form of wisdom through the ambiguous relationship maintained with the disturbed meaning/feeling of the states of awakening and dreaming. Science became recreational and fun thanks to the magic lanterns of the 18th and 19th Centuries which were used as a teaching aid. They are also used to better understand different scales of life through the magnified projections of microscopic animalcules. They are used as drawing machines and are described in science.

Without modestly dwelling on licentious plates used by the British prostitutes or at the court of Louis XV, it can be noted that the propagandist use of the magic lantern is also remarkable. The Jesuit Kircher proposed using it for religious preaching and the propagation of the faith. "We can demonstrate what we want," Kircher said; projection can be used as a political and social instrument or as an instrument for building identity. To a completely different degree, it was also used for military purposes, particularly during the French Revolution, when projectionist revolutionaries used lanterns to frighten nobles who were being pursued at night in the woods. It was, in this regard, an Austrian with a dual qualification as a military artificer and image projectionist named Uchatius who would be the first to couple a projector and an animated optical disc in 1853 with his *kinetoscope*. This one would then be part, in addition to the lineage of spectacular fireworks, of *Greek fire* projectors and *fire hoses*, and like mirrors that became "ardent" by Archimedes' trick that made them become legendary weapons. It is not unimaginable that the witches' bonfires during the Inquisition needed some optical devices to legitimize themselves through projections of filthy beings coming out of the fumes emanating from the burning evil bodies. The authors of the 18th Century phantasmagoria criticize their "necromancer" predecessors who were called charlatans and would abuse the credulity of the spectators. Inasmuch as is not the walking lanternists who are accused of pick pocketing in the 19th Century.

1.6. A figure of transgression and juxtaposition with a beyond

1.6.1. *Unconditionality*

There is, however, a use that does not find a temporal framework once we recognize what Kircher called the *summo stupore*: the visual devices that translate the philosophical traditions of illusion and that were used for different purposes all ultimately present a transgressive process that aims to see a *beyond* causing the "greatest astonishment". And, according to one hypothesis, it's probably one of their most timeless features. The desire to innovate, to experiment with new forms and to see oneself as a pioneer has a lot to do with the development of projection mapping. In the late Middle Ages in the 14th Century, Nicole Oresme acknowledged the pleasure of research and discovery in the face of knowledge which has come to a standstill, arguing that it would be better "if something always remained hidden, so that it could be the subject of future investigations". Any attempt at rationalization would itself lead to a chaotic emergence or resistance of the incoherent forms driven out by the same rationalization, surely? A primitive and ritual suspension of the world order could thus be at the root of the video projections transforming the gaze and the expected. At least if we agree to consider projection mapping as a medium that opposes the norm thanks to its ability to shape the pre-existing signifying perception. It will then be placed in the extension of the romanticism movement which is seen by some as a reaction to the philosophies of the Enlightenment that have failed in the attempt to access the ultimate truth and which aims to rehabilitate the irrational as a means of complementary knowledge.

The 17th Century saw the emergence of various attempts at global systematization at the risk of standardization. But when we chase away the representations of monsters in the so-called major art paintings of this classical period, we find them in the so-called minor framing and arts. It was in this context that Pierre Corneille became interested in the mechanical scenographies of the *Pièces à machines*, which allowed him to escape the control of the theatre practices imposed under Richelieu. This invites us to present projection mapping as part of the forms of artistic renewal that artists will seek and develop to escape the formalized and unsupported codes of their time. Another example: the grotesque Rabelaisian develops in an eschatological context. Thus, despite all the historical investigations undertaken in this text, we could wonder what an art without tradition would represent as a field of freedom! Walking in the green grass of the nearby garden, which no one can claim as his own, is always an ideal.

1.6.2. Magic image imagery

But is it always a question of fleeing from a world that is too bland, banal or suffocating with rules, or on the other hand, too little or badly stabilized? It would be wrong to stop at an exclusively emancipatory assessment of "chaos". It is understood that it may be at the opposite end of the spectrum to formalized codes by rejecting the standard, to see itself as a true superior order. Cultures thus value the "world of dreams", which is difficult to control and considered as important as that of awakening. The most famous on this subject are probably the Australian aborigines.

Since Mircea Eliade, shamanism has been recognized as an original form of religion based on a central pillar, namely the belief in a "world-other" inhabited by "spirits". Its access is reserved for initiated characters who can ensure intercession with the "common world". The shaman embodies the figure of transgression and this resilient practice is combined with contemporary and vernacular elements to the point of making historical and geographical specific characteristics a neutralized material. In the Middle Ages, for example, ancient Celtic traditions were extended and disseminated in the 13th Century in particular: they attest that the afterlife is among us, that the space of the dead coexists with that of the living and that both are intertwined despite appearances.

Divinatory practices using the "vision" are therefore also Shaman-like, this term being understood here in the double sense of ocular vision as well as ecstatic vision. Crystal balls, coffee grounds, screen-reading that enable us to read the future in the water or oil of a dish, extispicin or hepatoscopy looking at the entrails of animals are all sacred spaces of revelations and screen protoforms. Some of these can be likened to pareidolias. The Etruscan haruspices looked through the butt of a stick called the *lituus* to interpret the content of the gaze and thus deliver omens. This *templum*, framing a portion of celestial space, would give its name to the sanctuary of the temple that projects it onto the ground. Much later, in the 19th Century, the emerging possibility of remote communication with power tools led some to believe that it would be possible to communicate with the dead using these same tools. Technological spiritism then developed the fashion of magnetism and hypnosis in 18th Century Europe and gave itself the appearance of scientificity by taking inspiration from the establishment of telecommunication networks. These spiritualist phantasmagorias seemed to be accredited in 1895 by the already mentioned fortuitous invention of X-rays revealing the invisible bones of the living.

Seeing living skeletons does not fail to remind us of the phantasmagoria of the 18th Century, but we must also not forget that light already makes objects visible and recall that contemplating light in medieval times is an experience of Truth. In Strasbourg, one of the stained-glass windows projects a green ray precisely on the crucifix of the flesh at each equinox. The projection mapping that sublimates its medium in a luminous material is in harmony with the idea that any element perceptible in the Middle Ages is the luminous mark of an element to which Man cannot have direct access and implies the idea of (internal) harmony between things. The human spirit that offers itself to this radiance called *claritas* goes towards its transcendent cause, which is God. Father Suger was inspired by this anagogical approach in his conception of the cathedral by dazzling the senses of Christians with stained glass, candles, the sparkling golds and stones on which they are reflected. These lights were likely to cause ecstasy according to Suger. Erwin Panofsky would say that he prefigured the exhibitionism of today's film producer or fair organizer.

Luminous glass paste mosaics, some of which contained gold leaf, illuminations in manuscripts or religious icons are other luminous objects preceding the use of projectors. The analogy between projected pixels and the mosaic is quite obvious, but the analogy between projection and religious icons covered in part with gold may be less so. The icons would have given Véronique's first name from *vera icona* or the "true icon" since, according to the legend, she wiped the face of Christ with a cloth during the Way of the Cross. The icons then form the prototype of holy images from which the invisible but nevertheless luminous puts pressure on the visible.

1.6.3. *Anima*

The Greek word *anima* gives the root of the word animation. It is the art of making a soul or a spirit perceive in objects, drawings, why not groups or spaces: spirits of places or *Genius Loci* as the Romans called them. A German festival in Weimar bears this name and displays projection mapping. In the same category is the legend of the Golem: a mud being is animated by a phylactery on which the word "life" is written and placed in his mouth. The famous legend evokes the way in which an inert thing comes to life thanks to a code extract, just as the programmed digital computer image animates its medium. This legend can be found in the Bible during the creation of Adam and from the *Epic of Atra-hasis* or the *Poem of the Super-Wise* written in about the 18th Century BC.

The Greeks left several legends of animated artificial beings: the giant Talos and other statues of servants of Hephaestus, the sculptors Daedalus and Pygmalion giving life to their sculptures, and the legend of Pandora or that of Prometheus inspiring Frankenstein and Pinocchio. Today's generative electro-informatics systems are also digital environments that make it possible to cause things to emerge and create astonishment thanks to a behavior reminiscent of living things, beyond the mere automation of image creation that they optimize.

1.6.4. *See from a distance*

To close this mystical-magical field and its devices of vision of a *beyond*, it remains to tackle the suppression of distance. Lucien de Samosate, considered as the first author of science fiction, mentions the Selenite people in *Verae Historiae* who have existed since the 1st Century, and who have a universal sound and visual observation system. The latter consists of a well and a mirror above it. In *Alexander Romance*, the medieval bestseller probably written around the 2nd Century, the pharaoh Noctanebo could see the movement of the ships that came to attack him in a magic cauldron. He could also plunge figurines into the cauldron to sink the ones they represent *in real life* – this probably concerned lecanomancy.

However, lenses and their power to transform vision were known at that time. The oldest found to date is the *Bayard lens* dating from the 8th Century BC. Another one dated 79 BC was found at the Herculaneum in Pompeii. In the 20th Century, Ibn al Haytham spoke of the magnifying power of lenses. His work precedes that on the optics of Robert Grossetete or Roger Bacon which, in the 13th Century, described the telescope. The father and son Jansen would have built the first microscope in 1590. However, the telescope was not marketed until 1608 by Hans Lippershey before being made famous by Galileo in 1609. Would fantastic literature be inspired by it? *Le Berger extravagant*, which could be defined as the French equivalent of the Don Quixote written by Charles Sorel in 1627, presents two "magic mirrors" allowing one to see from a distance and to spy on the private life of his neighbors. He also describes surprising "recording sponges" of sounds. In the 18th Century, Charles-François Tiphaigne de La Roche wrote the novel *Giphantie* in which a people had a globular system with small imperceptible channels connected to the rest of the earth. You can listen and see universally thanks to a wand and a mirror reflecting portions of air reflected by spirits. Jacques Cazotte describes in the *inimitables Prouesses d'Ollivier Marquis d'Edresse* a character who helps the hero by showing him what is happening at Château de Tours at the bottom of a glass of water. Shortly afterwards, Goethe found a magic mirror (*Zauberspiegel*) in *Urfaust*. Then, in a 19th Century that still mixed technical patents and poetry, various authors imagined devices for transmitting images in fantastic stories: Charles Cross, Albert Robida, Edward Bellamy, Didier de Chousy, not to mention the famous Jules Verne.

1.7. The invention of an "empty box" as an image container

Among the optical devices of history, there is one that had a considerable influence on several others that preceded it, to the point that they were devalued or even made almost obsolete when they did not hybridize. This device is the screen, understood as a homogenized, quadrangular, flat and mobile opaque space. One could insist on the format or proportions that have become standardized. How old is it? The question could be just as thorny as the one about the origins of projection mapping, but we cannot avoid it if we want to understand how it can be considered as a specific medium that has been built or developed over time. It is now appropriate to focus on image carriers with a lack of meaning or expression when no image is apparent.

1.7.1. *Any precursors?*

In mortuary and religious art (altarpieces, stelae, tombs, etc.), in decorated "movable" objects such as tapestries and other decorated or painted sculptures (vases, masks, statues, pirogues, etc.), in architectural ornament known as "real estate" (frescoes, mosaics, bas-reliefs, stained glass windows, etc.), graphic compositions dress or disguise themselves by matching like costumes or masks the material and formal nature of their support, whose function always exceeds the reception of the image, whether they are painted, engraved or sculpted. We will probably find the first occurrences of the mediums strictly dedicated to visual content in the fabrics like banners, coats of arms and blazons. Or in tablets (clay, wax, stucco, wood, stone, etc.) or silks, papyrus, parchments, rolls, codices, books, cards, etc. Were these media nevertheless meaningful, like the Tuareg who use techniques related to the nature of the media to write, while they come from a choice in a range of materials defined by the meaning of the message? Has there been an emergence of what can be described as a neutral container with decontextualized content, such as the museum *white box* or the gallery criticized by *land artists* and conceptual artists?

1.7.2. *Alberti and the invention of the screen*

In the theory of the painter Leon Battista Alberti and his book *De Pictura* dated 1435, there is a description of a "first act" which consists in "drawing a rectangle, of the size that [he] is comfortable with as an open window through which [he] can see the subject". In the Albertian model, the world must be divided into a portion that can then be taken with you: the painting painted on a medium considered as a void. An "intersector veil" structured by a regular orthonormal grid is placed between the observer and what is represented on the board. Alberti reasons that perception is reduced to a pyramid, which reduces sight to a single Cyclops point from which all spatiality receives its determinations. With this observation point and the leakage points of the construction in perspective, it is abstract mathematical entities without surface or thickness that become both the origin and the horizon of the image. Thanks to this mathematical systematization, we obtain a representation that is intended to be faithful to what is perceived, since "what cannot be grasped by the eye has absolutely nothing to do with the painter". This sentence underlines that for Alberti, no ambition seeks to go beyond the visible. Alberti theorizes a type of image that deliberately ignores its medium by making a nothing or a hole which becomes an "open window". Visual heteronomy then gives way to the abolition of the expression of the medium and the perspective is then imposed in other genres: politics, science, literature, theatre, etc. It determines the relationship to authority and truth. The 15th Century, later described as "modern", produced the visual device that this modernity took as a model for seeing the world. The Albertian window becomes the dominant or even totalitarian visual construction. Until the questioning of this type of construction becomes the basis for a whole section of what will also be called, but in a self-proclaimed way this time – is it ironic? – "Modern art", from the end of the 19th Century and the beginning of the 20th Century.

1.7.3. *The humanistic context of the disruptive object-subject disconnect reified in and through the image*

Behind this somewhat abstruse subtitle lies a hypothetical deduction inspired by speeches by historians, anthropologists or sociologists, on the pivotal moment that saw the screen system unfold. The *modern* inflection in the course of history from the *quattrocento* allowing the Albertian model to develop invites us to consider projection mapping according to the axiological character of this period during which an original framework of thought was built and developed. This framing completes the pictorial framing, forcing nature to flow through an inflexible grid into a preformed conceptual box; Renaissance humanism, built in particular on a base of Greek aesthetics and hermetic esotericism and then marked by the philosophy of enlightenment, leads to a form of devaluation of *non-humans*. In the 16th or 18th Centuries, during the transition from a closed world to an infinite universe, Man came to consider himself as a center, no longer defining himself by links to space or to the objects that surround him. Then, after losing interest in its medium, the image excludes the subject himself on a Cartesian basis of rationality of the "real". Space or objects take place on the other side of a conceptual dividing line drawn between "subject and object" as Bruno Latour says, or between "nature and culture" in Philippe Descola's words. What is not considered human, whether it is of "nature" or artificial, is seen from the outside which tends to make it a simple source of potential and availability. The asserted rupture with a devalued past doubles this ideology which claims to be progressive. The psychological mode of expression of things is refused as old beliefs to be overcome. The lack of a historical perspective in favor of a predominant understanding of the space of so-called "traditional" cultures, for example, is seen by the West – seeking a pretext for colonization – as a feature of primitivism. This reading in turn generates the contemporary myth of the space devoured by time, or that of the mechanization and objectification of space in an exclusively quantified model that leads to the neutralization of perceptions since it refuses its heterogeneity by arguing the equivalence of the points of which it would be constituted. If the Albertian framework and the theory of perspective develop in this context of "separation" and feed it, today's screen has its origins there and projection mapping would be opposed to it.

1.7.4. *A hypothetical starting point*

Projection mapping could not therefore exist without the institution of the screen perspective which would give it a negative starting point ensuring it is positive. That is, if we agree to put aside what, before the Renaissance, leaves us to suppose predecessors to the Albertian model. The *emblemata* in ancient Greek mosaics are an example. Pliny the Elder also tells us in the 1st Century that the only real painting was wood paintings, and that it had almost disappeared in his time in favor of mural paintings. The well and the mirror of Lucien de Samosate's *Vera historiae* or even the aruspices stick could in turn challenge the idea of an Albertian revolution based on a symbolic disregard for the nature or shape of the painting canvas. It must also be admitted that a totally decontextualized media remains difficult to define theoretically: can an image do without a corresponding surface or a consideration of its space? If it must necessarily agree, this would suggest by pushing the reasoning that projection mapping is nothing more or less than a pleonasm playing the game of an imposture characteristic insofar as everything could be considered as contextual and relational, therefore using "correspondence"! Rem Koolhaas' caricatural aphorism *fuck the context* probably sums up that such an idea is not conceivable. Without further evoking the shared intuition that not every projected image is necessarily projection mapping: what remains to be determined if we want to define clear outlines?

1.8. Modern inflexions: obsolescence of old visual devices and tacit challenges to the Albertian model

We could therefore assume as a prerequisite to defining projection mapping that there is, on the one hand, a visual object that is opaque, closed, folded up on itself, and on the other hand, a transparent, relational or heteronomical object. It is now the dynamics of mutual influences of these two forms that must be observed in terms of the history of their development.

1.8.1. *Obsolescence*

According to the medievalist Paul Zumthor, manuscript painting in the Middle Ages remained "unframed" by remaining "on the same level as the surface on which it was inscribed and could only be understood (even when a decorated line encircled it) in relation to it". The "under-frame" painting, he argues, which covers a visibly delimited geometric surface and separated from the surrounding space by a line, a border or any other linear boundary, will realize its latent values with easel painting in the 14th Century.

It was no different with the wall paintings that punctuate here and there with a significant difference in the regular quadrangles of the frescoes. After having been one of the most practiced modes of expression from Antiquity to the 18th Century, with a passage through a 13th Century which was its most fertile period, the fresco was in turn abandoned in favor of easel painting. From the 17th Century onwards, monumental wall decorations were preferably made on panels or large canvases in frames. The realists and impressionists did not accept this medium, which regained interest in the 20th Century with Masson (Odeon), Chagall (Garnier opera), and Braque (Louvre).

The same applies to sculpture in the West in the 11th and 12th Centuries, which is integrated into the architectural structure of religious buildings. Details and characters are strictly subordinate to its primacy. It took four or five centuries, from the 11th to the 15th Century, to ensure that the sculpted figures had their full tridimensionality, the substantiality of their own space. The sculptor's work will only really break the ties that bind him to his background in the 16th Century. What the base will provide.

It was also at Christmas 1492 that the first confined space theatrical performance took place, to the detriment of processions and outdoor demonstrations. The live show is then transformed into an animated painting. The mosaic, on the other hand, declined after the capture of Constantinople in 1453 by the Ottomans. As for stained glass, while it became grey in the 14th Century under the influence of Cistercian and Franciscan thought, the search for clarity from the classical period to the 17th and 18th Centuries would lead to its decline after a peak in the 13th Century with the development of the Gothic style and its large bay windows.

1.8.2. Challenges

Nonetheless, in a movement that goes beyond Italy alone, mannerism in the 16th Century played very early on with Albertian landmarks in a set of codes and symbols that are often murky, mixing borrowings, quotations and distances with the new humanism. By producing new emotional and artistic effects, this movement deliberately sought to break with the exact proportions or reality of this new Albertian space. Still famous are the *Self-portrait in a Convex Mirror* by Parmigianino (1524) or *The Ambassadors* by Hans Holbein the Younger (1533) which depicts a homographic-type anamorphosis hiding in the foreground, like those commonly used in projection mapping. The meaning of these anamorphoses is probably different from those theorized in the 17th Century by Jean-Louis Vaulezard, for example, which aimed more at highlighting the inaccuracy of the human gaze than at questioning the relationship to authority and truth that the central perspective establishes or supports. From mannerism, the medium no longer seems to make sense beyond its dimensions, except perhaps in still life without great depths that will open the way to trompe-l'oeil where play then fully assumes itself with the medium on which the image is presented. Take, for example, the *Reverse of a Frame Painting* by Cornelis Norbertus Gysbrechts. In Tuscany, from the 12th to the 14th Centuries, the council chambers of municipal palaces and private palaces were enriched by numerous decorative cycles. Giotto provided some of the most beautiful frescoes of this time and the Mannerist movement brought a renewal by treating the frescoes for themselves. By turning towards an unrealistic art such as that of Pontormo, he offered the most elaborate examples of compositions where separately applied paintings, false columns, niches and fictitious openings blended together. In the 18th Century, work on Father Andrea Pozzo's perspective and trompe-l'oeil influenced the whole of Europe. Painting now gave the illusion of sculpture and architecture.

Yet in Baroque, Classicism, Rococo, Neoclassicism, Romanticism, Naturalism, Realism, Pre-Raphaelitism, etc., as well as in Symbolism and Impressionism, framed painting is a reference. From the 17th to the 18th Centuries, painters such as Piranese or Turner brought down the perspective before the Viennese secessionists from Austria at the end of the 19th Century, such as Gustav Klimt and the Symbolist movement with Jean Moréas, reacting against naturalism and academism. In an art that moves away from the description of the material world to devote itself to the sensitive representation of the *Idea* through words, images or sounds, symbolism engages in such a project that it goes hand in hand with a need to decompartmentalize the performing arts that it strongly marks at the turn of the century. The modernity of Cézanne or van Gogh, of impressionism, then changed the artistic scheme so that the coexistence of the various vision regimes of the 20th Century and the various transgressions that followed would take place. Sublimation of matter from a Hegelian perspective on the one hand (Kandinsky, Malevitch, Moholy-Nagy) or materialistic valorization on the other (Masson, Tinguely, Gutaï). Extension of the limits of the table and accession to new dimensions (Tatline, De Stijl, Lissitzky). Fusion in electro-mechanized (Loïe Fuller, futurism). Synesthetic art in order to find a universal language (Bauhaus). Or the search for a *totality* to the point of flirting with the common banal, at the price perhaps of one of the specific features of art. These aesthetic and plastic investigations are as many questions about the Albertian window as we could place in the family tree of projection mapping. Many painters have taken up the medium of animation in order to give their paintings the dimension of time (Léopold Survage, Walter Ruttmann, Hans Richter, Berthold Bartosch, Alexandre Alexeieff, Lotte Reiniger, Oskar Fischinger, Viking Eggeling, Stan Brakhage, Man Ray, Len Lye, etc.) and many stage directors have aimed to give space a plastic presence and autonomy vis-à-vis the text (Adolphe Appia, Caspar Neher, Jo Mielziner, Christian Bérard, Josef Svoboda, Ezio Frigerio, André Acquart, Oskar Schlemmer, Alwin Nikolais, Vsevolod Meyerhold, Edward Gordon Craig, Tadeusz Kantor, etc.).

Just as the details of the 18th Century projective systems that were scientifically described in academies and used as machines to automate drawing or as a pedagogical support were kept aside, it is unfortunately not possible to immerse oneself more in the history of modern and contemporary art here if we want to achieve a reasonably sized text.

1.9. Parastatic scenography

Part of the art history that followed the Renaissance, which began with mannerism, can be read through the prism of the questioning of the codes that Alberti theorized. At the same time, its new screen is spreading in many fields of modern society. The impact is such that other art forms are gradually transformed or lose interest: mosaics, stained glass, wall or manuscript painting, sculpture. The theatre, for its part, locks itself inside and puts itself in perspective, to become the equivalent of a painting, although animated and lively.

1.9.1. *For the eyes: the uncomplicated image*

The words video, projection and mapping do not reveal what the discipline of projection mapping owes to the worlds of entertainment and scenography, which today use it as part of an ancient tradition of dramatic effects. To illustrate the ride of the tetralogy *Der Ring des Nibelungen* in 1876 in Bayreuth, the painter Carl-Emil Doepler, responsible for the costumes, suggested painting glass discs to be projected using a magic lantern. In Paris, they are clouds projected from a transparent disc. The poet Aeschylus, born around 525 BC in Eleusis, where Matt Gatton imagines statues animated by *camera obscura* projections, already introduced the idea of the *Deus Ex Machina* bringing the gods by mechanisms. However, it was in the 16th and 17th Centuries that the original form of the *Pièce à machine* was invented. Pierre Corneille finds with this theatrical type a good alternative to the politically imposed constraints: he wrote in 1650 in *Argument d'Andromède*: "My main goal here was to satisfy the view with the brilliance and diversity of the show, and not to touch the mind with the force of reasoning, or the heart with the delicacy of passion. ... this show is only for the eyes."

The expression "sound and light" is sometimes used today as a clumsy synonym originating from mannerism and is not hesitant to compare it to a renewal of traditional fireworks. Spectacular pyrotechnics appeared in the West at the end of the 16th Century in England and in the 17th Century in France. In the 18th Century, Handel caused traffic jams on carriages for fire and music. In a field close to these sets of lumino-chromatic and sound correspondences, other artists have interpreted or imagined colored music that prefigures the sublimation of the material medium of certain projection mapping into an intangible visual sound: the mannerist Arcimboldo, followed in the 16th and 18th Centuries by Louis Bertrand Castel, Philippe Rameau, Philipp Telemann, Karl von Eckartshausen, Johann Gottlob Krüger, the Kircher harmony organ, not to be confused with the cat organ, etc. The study of synesthetic experiments continued in the 19th Century with various devices: the psychedelic *kaleidoscope* which became *kaleidophone*, or the *chromatrope of* the 1850s. Thanks to its crank handle, it rotates two metal crowns in opposite directions and projects abstract psychedelic images. It would be followed by *Astrometeoroscope*, *Eidotrope*, *Kaleidotrope*, *Cycloidotrope*, etc. The *Liquid light show* of the 1970s then marked the *New Age* period.

1.9.2. *Living presences and images*

Sebastiano Serlio, born in 1475, is said to have been the first to use the word scenography. From the 16th to the 18th Century, Niccolo Sabbatini, Giacomo Torelli and Ferdinando Galli da Bibiena were names that marked this art form in its early days. Galli da Bibiena designs sets whose piranesian vehemence proposes vertiginous abysses, in infinite corridors that raise doubts about a total mimetic subjection to an Albertian rational perception. On stage follow the famous phantasmagoria of the 18th Century, in which the dead appear as skeletons or ghosts. They are part of a long tradition of thaumaturgic performance, including Chinese shadows that are at least 15 centuries old. François Dominique Seraphin, around 1772, is said to be one of the first to produce this type of "Chinese shadow" show in France. Félicien Trewey between 19th and 20th Centuries would follow the same tradition.

After the introduction of shadow shows by Seraphin, the escape point immersed in infinite optical feedback proposed by Galli da Bibiena would turn like a glove from the 1780s onwards into immersive panoramas. They updated the large format architectural ornament of the spatialized visual sequences contained in the tapestries, mosaics, wall paintings, sculptures or low reliefs mentioned above in a radically concave and curly form. In 1781, Philippe-Jacques de Loutherbourg animated what are still called panoramas, but they were framed and in a more modest format: paintings were presented successively accompanied by plays of light such as the small mechanical theatre of mirrors and pulleys that he called *Eidophusikon*. Louis Daguerre, in 1822, created the "diorama" which developed this principle in a show of skillful combination of painting and lighting. They produce a striking illusion on the viewer. Other "myorama" or "moving panorama" were produced between 1810 and 1880. John Banvard, with his "georama", and Moses Gompertz remain prominent names for these scrolling paintings, as is the *Grand Moving Panorama Of Pilgrim's Progress* of 1850, which is 800 feet long by 8 feet high.

In the engravings that represent these moving panoramas, we notice men who manipulate the images in public view or who bring a vocal complement. This was also done in Japanese cinemas from the early days of cinema until the 1930s with the *Benshis*. The latter commented, repeated the dialogues, and read the title cards to the illiterate public. The well-known and acclaimed benshis attracted more crowds than the actors, directors or films themselves. As early as the 18th Century, the great organizer of the Duc d'Orléans' celebrations, Louis Carrogis, known as Louis de Carmontelle, played improvised and interactive comedies before taking his characters for a walk in projections from transparencies. We can also recall that until about 1907, the system of representation of "primitive" cinema was based on the presence of actors. They performed for an audience as if they were in the theatre and the style was strictly frontal. At that time, the space between the cinema and the screen was so separated that viewers could react, come and go while keeping a psychological distance from the cinematographic narrative.

In the same vein, the word *cartoon*, which means "caricature", reminds us that many of the animators at the beginning of the cartoon come from the press cartoon: Georges Méliès, James Stuart Blackton, Émile Cohl, and others. The press demands a clear line and a speed of execution that has allowed them to weave a link between animation and the world of live performance to give the *chalk-talk* or "talk with a chalk" and the *lights sketches* or "flash drawings". During these shows, a cartoonist on stage caricatured people on a blackboard with chalk. Winsor McCay performed on stage in vaudeville like this and invented *Gertie*, a dinosaur projected on a scale of 1:1 with which he showed himself in a trick training sequence.

1.9.3. *From the screen to film*

Although Émile Reynaud invented "optical theatre" in 1889 and its multiple manipulations of images that are performative in nature, he did not inspire many of the animators who would follow. The latter are more influenced by *flip-books*, which are more widely distributed. These small objects, also called "folioscopes", extend a line of screen objects. Smaller in size than the large moving panoramas, Chinese illuminated scrolls were imported to Japan in the 6th Century. They take the name of *emaki*. Long strip of 50 centimeters high and proto-animate which are then unrolled, they gave the famous scrolls of the *Qing Ming Ming Festival*, that of the *Birds and Animals* or that of *Grand Councilor Tomo no Yoshio* in the 13th Century during the Song dynasty.

Another plausible ancestor of the *flip-book* is the book of Heidelberg illuminations painted in 1558: it preserves an identical framing throughout its pages. The sequences of narrative images then took the form of discs of optical toys: the "thaumatrope" (John Hershel and John Ayrton Paris 1825), "phenakistoscope" (Joseph Plateau 1832), "zootrope" (William George Horner 1834), *flip-book* or "flipbook" (Pierre-Hubert Desvignes around 1860), "kineograph" (John Barnes Linnett 1866), "chorentoscope" (Lionel Beale 1866), or finally the "praxinoscope" (Émile Reynaud 1876) to which Émile Reynaud added a projector in 1880.

These images became the film space in the cinema. But after about 1907, this space no longer functioned as a backdrop for the theatre. The viewer was then placed in the fictional universe of the story: he or she is asked to identify with the characters by living the story from their points of view, having the best possible distance and angle in each shot. The spectator then finds himself inside a space that does not really exist, or rather that no longer communicates with reality, but becomes so through a skillful play of variations in the frame and the temporal montage of images. Paradoxically enough, however, the editing theories of the Russians Koulechov and Eisenstein in the 1920s were influenced by dance for the former and probably by the stage director Meyerhold for the latter. The living spectacle, problematizing in a living way the relationships between humans, had begun to question and integrate within itself what the techniques generate in these relationships. The cinema projector has not escaped this; and as a fair return, the fixed shot theatrical form abandoned by cinema would see a return 100 years later with projection mapping, whose material device, however, differs very little from that of cinema.

1.10. From expedition to investigation

1.10.1. *Resilience*

Projection mapping often bears the hallmark of innovation and technological progress. It is sometimes used to stage the idea of modernity or a vision of the future, and it does so in contrast to traditional screen writing such as cinema, television, or computer and phone screens. However, these latter can be seen as more recent devices than the projection mapping device. If it is sometimes described as avant-garde art or exposed to stage rhetoric of openness to the future, would it be described as hyper-modern, alter-modern or pre-modern when combined with a historical study? Is projection mapping an innovation or a redesign of a type of image that had been devalued? From our expedition, we learned two contradictory things and tried to solve this problem by presenting a figure of timeless transgression. This recognizes both the historical continuity of a practice through the renewed rupture that allows otherness and astonishment.

Would projection mapping be able to synthesize the Albertian window with the above? On the one hand, it contests the Albertian construction in its exclusive graphic elaborations while refusing, in favor of visual heteronomy, the autonomy of an image inaugurated in the modern paradigm of nature/culture or object/subject divide in the West. On the other hand, however, it also extends this construction by being part of a contemporary proliferation of screens and by using the perspectivist theory used in some trompe-l'oeil plays.

Looking too much at changes, evolutions or differences, the risk would be to neglect invariants: it is perhaps not the many innovations that matter in history, but why one or the other has managed to maintain prominence and how it has spread. Projection mapping confuses its origin with that of known pictorial representations and with the human predisposition to transform bodies. It is part of the traditional ritual space-time inversion to be observed in the light of post-colonial anthropology. It would therefore be a living and resilient form that suggests that by repeating itself it has returned in an original form, or will do so again in the future. Following this hypothesis, projection mapping would be a cross actualization with the digital and electronic technologies of the timeless spectrum of the median and moving figure that marks the passages from interiority to exteriority and vice versa. This figure, nourished by the field of tension caused by a desire for emancipation that is sometimes incompatible with attachments that contrast with it, has never ceased to produce hybrids, of which projection mapping is a part, as soon as it is studied in its aesthetic dimensions. And projection mapping would then in turn configure the sensitive experience, which remains to be studied.

1.10.2. *Ongoing investigation*

After moving from X-rays or micro/tele/scopes lenses to the automation of figurative image production, from the worlds of personal or collective dreams to data visualization or science didactics, it emerged from the understanding of this collection that projection mapping is more than an expression tool and can also be considered as a visual device that has developed, and which promises in this form to show new things. It now requires a detailed and methodical analysis beyond this speculative basis.

This chapter of history did not cover ethnographic discourses, nor the development of projection mapping from the perspective of the fairground equipment, which, nonetheless, as with cinema, was of considerable importance. Many details on the movements of modern art during the transition to the 20th Century are also lacking, as well as the so-called post-modern movements from the 1960s onwards such as land art, street art, contextual art, the psychogeography of situationism, the art of installation, etc., which nevertheless find their place in a projection mapping genealogy. We also did not focus on the encounters between electronic devices and artists starting from the same period: cybernetic art, video art, digital scenography, etc. Finally, we would benefit from retracing the way in which the expressions "video" or "projection mapping" were claimed after the 2000s, in particular through tutorials, *software* or *hardware* tools, distribution spaces and creative bodies. We therefore leave the matter, which questions the reasons for the rise of "video mapping", open as of 2008.

In addition, from a desired second part to this historical introduction, beginning in the 20th Century, the three axes envisaged for the extension of our work are as follows: first, to identify and define as best as possible what projection mapping is in terms of the realities it covers today. Second, describe what it consists of and how it is produced from a systemic perspective. Third, what are and how can we describe the original phenomena it produces? The contributions to be followed in this book cover some of the issues raised by these three axes.

1.11. Conclusion

To approach the answer to the question of the origin of projection mapping one last time, I will confine myself to repeating what the new media theorist Lev Manovich noted, for whom the industrial production mode would have meant that cinema has replaced other spatialized narrative modes with a sequential narrative. It is the fact of obeying the logic of breaking down a task into a series of elementary operations performed one by one, says Manovich, that has meant that the coexistence of images has not been systematically explored from the very beginning of cinema, with a few notable exceptions, such as the *splitscreen* used by Abel Gance, the *expanded cinema movement* of the 1960s or Émile Radok's *diapolyecran*, on display at the Czech pavilion of the 1967 World Expo. With Edward Soya, Manovitch notes that "the decline in the imaginary conception of space and the spatial conception of social analysis coincides with the rise of the paradigm of historical consciousness in the second half of the 19th Century. It was only with the rise of ideas such as "geopolitics" and "globalization" that these approaches were able to return during the last decades of the 20th Century". But with the computer and screen as it has been designed since 1970, each pixel corresponds to a memory box that must be activated by the program. The logic of a single image/screen breaks down and the computer interface then begins to use multiple visualization windows from which Manovich draws perspectives that we can take for ourselves: "We can logically expect computer-assisted cinema to follow this path", and "when it happens, the tradition of spatial narrative that 20th Century cinema had suppressed will be reborn".

1.12. References

Azéma, M. (2011). *La préhistoire du cinéma: Origines paléolithiques de la narration graphique et du cinématographe*. Errance, Paris.

Barbanti, R. (2009). *Les origines des arts multimédias. L'influence des mnémo-télé-technologies acoustiques sur l'art*. Eyrolles, Paris.

Beaune, J.-C. (1992). *L'automate et ses mobiles*. Flammarion, Paris.

Burczykowski, L. (2014). Par-delà l'écran. PhD thesis, Université Paris VIII, Saint-Denis.

Carbone, M., Bodini, J. (eds) (2016). *Voir selon les écrans, penser selon les écrans*. Mimesis, Sesto San Giovanni.

Denis, S. (2011). *Le cinéma d'animation*. Armand Colin, Paris.

Descola, P. (2005). *Par-delà nature et culture*. Gallimard, Paris.

Garbagnati, L., Morelli, P. (2006). *Théâtre et nouvelles technologies*. EUD, Dijon, p. 81.

Gatton, M., Carreon, L., Cawein, M., Brock, W., Scott, V. (2010). The Camera Obscura and the Origin of Art: The case for image projection in the Paleolithic. In *Official Proceedings of the XV World Congress of the Union internationale des sciences préhistoriques et protohistoriques*. Archaeopress, Oxford.

Gell, A. (2009). *L'art et ses agents. Une théorie anthropologique*. Les Presses du réel, Dijon.

Katz, S. (2004). *L'écran, de l'icône au virtuel: la résistance de l'infigurable*. L'Harmattan, Paris.

Lange, A. (2000). Le miroir magique. La vision à distance par le biais de l'électricité à la fn du XIXe siècle et l'historiographie des origines de la télévision. *Recherches en Communication*, 14, 53–72.

Latour, B. (2006). *Nous n'avons jamais été modernes. Essai d'anthropologie symétrique*. La Découverte, Paris.

Lefrère, J.-J., David, B. (2013). *La plus vieille énigme de l'humanité*. Fayard, Paris.

de Loisy, J., Hell, B. (eds) (2012). *Les maîtres du désordre*. Quai Branly/ RMN, Paris.

Maitte, B. (2015). *Une histoire de la lumière. De Platon au photon*. Le Seuil, Paris.

Manovich, L. (2010). *Le langage des nouveaux médias*. Les Presses du réel, Dijon.

de Mèredieu, F. (2008). *Histoire matérielle et immatérielle de l'art moderne & contemporain*. Larousse, Paris.

Perriault, J. (1996). Les simulacres de lumière: une archéologie. *Les cahiers de médiologie – La querelle du spectacle*, 1, 49–57.

Picon-Vallin, B. (1998). *Les écrans sur la scène: Tentations et résistances de la scène face aux images*. L'Âge d'Homme, Lausanne.

Sconce, J. (2000). *Haunted Media. Electronic Presence from Telegraphy to Television*. Duke University Press, Durham.

Yates, F. (1987). *L'art de la mémoire*. Gallimard, Paris.

Zumthor, P. (2014). *La mesure du monde. Représentations de l'espace au Moyen Âge*. Le Seuil, Paris.

Webography

History of the discovery of cinematography:

History of LEITZ projectors:

The adventure of writing:

2

The "Spatialization" of the Gaze with the Projection Mapping Dispositive

"In another moment, Alice was through the glass, and had jumped lightly down into the Look-glass room. The very first thing she did was to look whether there was a fire in the fireplace, and she was quite pleased to find there was a real one, blazing away brightly as the one she had left behind. 'So I shall be as warm here as I was in the old room,' thought Alice; 'warmer, in fact, because there'll be no one here to scold me away from the fire. Oh, what fun it'll be, when they see me through the glass in here, and can't get at me!'

Then she began to look about, and noticed that what could be seen from the old room was quite common and uninteresting, but that all the rest was as different as possible. For instance, the pictures on the wall next the fire seemed to be all alive, and the very clock on the chimney-piece (you know you can only see the back of it in the Look-glass) had got the face of a little old man, and grinned at her."[1]

2.1. Introduction

The Mirror as a metaphor for mirrored, reflected, mapped reality is one of the most widely used images in many human sciences today. It is from this

---

Chapter written by Justyna Weronika ŁABĄDŹ.
1 Lewis Carroll, *Through the Looking Glass*, 1871.

image that the 19th Century English writer Lewis Carroll created his famous novel, the story of Alice, entitled *Through the Looking Glass*. When she passes through the mirror, that is, when she passes from reality to its reflection and image, Alice draws conclusions that recall general observations about illusions created by imitating reality. These concepts have been developed and evolved over the years in painting, photography and, finally, in cinema and its audio-visual derivatives. The fireplace on the other side of the mirror attracts Alice even more – although it is the same as the one on the side she just left, the fire is even hotter, because there will be no one to scold the girl if she approaches it. Everything seems identical and uninteresting, but some of the objects in the room, such as the paintings above the fireplace or the pendulum, suddenly seem to have another life: a magical life. It is therefore understandable that this imagined reality can be even more attractive. Not only does it remind us of the one we know, in which we live, but for us, the public, it also removes the limits that existed before and prevented us from getting close enough to it to feel it, to experience it. In this imaginary world, the limits of what can be done in this "reality" are also removed. Our imagination seems to be the only real limit. The purpose of this quotation is not so much to explain the attraction of modern spectators to imaginary, illusory reality as to highlight the changes in perception and the evolution of needs in the lives of people who have appeared with the invention of audio-visual forms – from simple projections with the darkroom, phantasmagoria, magic lanterns, to modern forms and still developing virtual and augmented reality, through cinema. In realizations based on virtual or augmented reality technologies, the light of the "source of fire" is also more attractive than the one we encounter in our daily reality. It allows us not only to see people and all the other forms that make up our world, but also to create illusions, simulations and visualizations of things that are not real. The optical illusions that create the imaginary worlds of film and other audio-visual forms, including increasingly immersive forms of augmented or virtual reality, remain the main attractions for viewers fascinated by their power of creativity and immersion. This immersion is such that, like Alice, the spectators and participants of these audio-visual events pass through the mirror, and the reality they find there amazes them with the possibilities it offers.

2.2. The release of the "cinematographic cocoon"

Projection mapping, also known as video or 3D mapping, is part of the family of illusion-based audio-visual forms, which today transform not only the way animation is perceived, but also the perception of viewers. After the post-production process, computer-generated 2D and 3D animations (and synchronized with sound) are projected onto previously measured and studied objects that become screens. These forms of audio-visual art thus allow the creation of a new augmented reality within the reality in which we live, which together form a hybrid. Without giving up their prototype, however, they are freed from their material reality as it was known until then. Using, or even from digital animations, they usually create a simulation of a world that has no connection to reality. According to Jean-Louis Baudry's psychoanalytical theories, looking at them in a crowd (as is the case with architectural mapping) transforms the viewer's perception into a virtual hallucination that everyone experiences during the show (Baudry 1992, p. 87). Reality simulations in the form of projection mapping also tend towards the development of visions and concepts of sign creation that no longer represent or even simulate anything and focus on the ontological confrontation of the similar with the similar as the critic Jean Baudrillard notes (Baudrillard 1983, pp. 70–98). Apart from this resolutely critical attitude of the French cultural sociologist and philosopher towards virtual reality technologies, projection mapping can be presented as an extension of expanded cinema, a form described in 1970 by Gene Youngblood, who referred to many experimental cinematographic works (Youngblood 1970). Projection mapping goes beyond the boundaries of the screen itself, because any object can become a screen – from the smallest objects to the colossal architecture of buildings of several square meters (as in the case of the projection mapping of the Parliament of Bucharest, Romania). Its dimensions are only limited by the technology and budget available.

Borders are also being crossed thanks to the modification and expansion of the possibilities of the viewer's perception of the audio-visual image. Projection mapping viewers are freed from what Roland Barthes calls the black of the "cinematographic cocoon" (Barthes 1993, pp. 157–160) that envelops them when they immerse themselves in the images, sound and history of the film in a specially adapted place immersed in the dark: the cinema. Projection mapping animations also need dark spaces. This is why they are usually presented in previously adapted places with limited light, or in natural darkness, after sunset.

Often, these presentations require additional authorization to turn off the city's lights. As in a cinema, the dark space of projection mapping makes it possible to concentrate on the flickering of projections, on the illusions that move objects that were previously stable and immobile. Symbolically, the spectators of a cinema sitting face-to-face with the screen become the unique receivers of the images. During the projection, they no longer have any contact with external reality, which allows them to experience "narcissistic identification", that is they identify themselves with the perceived images (Metz 1977). At the same time, with projection mapping, even if we generally participate in this grandiose spectacle of sound and light in communion with the crowd, this type of concentration is impossible. First of all, this is due to the fact that projection mapping shows are generally shorter animated image presentations than those we see in the cinema, where films last at least an hour. In addition, the spectator is generally not seated in a comfortable seat, especially when it comes to installations using architectural mapping and mapping. Its movements are therefore not restricted, and his attention is not reinforced by the immobility of his body on the seat. It can move freely and even change places, which is necessary for some installations. An immersion similar to that of cinema will be more possible in stage mapping projects, which introduce projection mapping elements into plays or ballets. These projections serve as an alternative to scenography and allow even more illusory modifications of space and effects previously impossible in the theater.

Actors and dancers interact with the unreal, the imaginary reality opening up new possibilities on the theater stage. Founded in 1958, Laterna Magika's team, which since 2010 has been participating in the theater programme of the Národní Divadlo (National Theater) in Prague (Národní Divadlo on the Prague website), is an excellent example. If Laterna Magika's name refers to the most basic projection device created in the 18th Century, the work of this group has nothing to do with simple visual projections on slides. The team offers a wide range of multimedia shows that combine standard projection systems and projection mapping with the movements of actors and dancers on stage, creating exciting shows where the spectator is completely immersed. One example is *Cube*, a show directed by Pavel Knolle and first launched in 2017. The projection mapping created for this show was based on a form built for a specific purpose – a scene in the scene – the cube to which the actors dance. The stories they present are the result of the incredible combination of the beauty of moving bodies and their interaction with space, which, in this case, has been technologically designed and dynamically evolved. Sitting on the seats of the theater, perfectly concentrated, the spectators can totally lose themselves in these visions that change at full speed. In this way, technology frees the theater from its frontality and the limitation of perspective that has been imposed on the public until now. Thanks to illusion, the perspective changes and leads viewers to cinematographic-type creations. It is a symbolic upheaval for modern man, suspended between the tangible, the real reality and the reality created by digital technology that expands the capacities of our body, here and now.

Figure 2.1. *Laterna Magika, Cube, director Pavel Knolle, National Theater of Prague, Czech Republic (source: photograph by Benjamin Peguero). For a color version of the figures in this book see, www.iste.co.uk/schmitt/image.zip*

2.3. Changing the projection mapping dispositive

The term "dispositive" is one of the most often misunderstood and misused terms in the research methodology of film studies. It is not the name itself, often limited to its French form, "*dispositif*", which has become very popular, or in rare cases translated into the target languages (for example *dyspozytyw* in Polish). The problem arises first of all when it is used to describe the mechanism of media-related phenomena related to perception. Moreover, its meaning is often limited to a device, which does not allow the whole character of the phenomenon to be reflected. The term was borrowed, in a rather unfortunate way, from Michel Foucault, who used it to describe different social practices codified as rules and codes, and other conventions in the form of an organization of space and norms of conduct, which reflect the social relations of power and knowledge.

Initially, Foucault had developed his theory as part of his *Histoire de la sexualité*, but in analyzing this work, the French philosopher Gilles Deleuze attributed another meaning to the term "*dispositif*" (Foucault did not invent this neologism, it is a term that already existed in French). Deleuze notes that visibilities are only the ways in which we see the objects and results of our senses, but are created by epistemological formations that constitute a "dispositive" of light (Deleuze 1986, pp. 69–98). The spectators, therefore, as a minor element of these visibilities, but also of the machines that create them, allow them to exist and function. This "vision machine" was the cinema and its various optical devices. In the context of film and media research, the "dispositive" is therefore the order of reception, which includes and studies an audio-visual work in different media contexts. These contexts consist of a spectator, a space and a medium (Gwóźdź 2003, p. 38).

Projection mapping will use devices, machines and, therefore, a slightly different "dispositive" than cinema. It adopts various forms of projection, placed in different contexts. First, it is based on the removal of the flat, rectangular screen that restricts and limits the spatial possibilities of the film. From now on, any object can become a screen. These are primarily three-dimensional objects. The stories created using projections designed for this purpose are part of the object's tangible volumetric structure, which literally (and not by illusion) projects towards the public. Just as modern cinema no longer needs film, images are projected with a digital projector that uses computers and mathematical operations to transform binary code into virtual images. This image, like other visual compositions, has the following characteristics: color, chromaticity, creative ability to use the natural elements of the object and, like film and animation, to constantly modify them to create an illusion of movement. Its parameters, such as lumens (lm) or luminous flux units that determine the amount of light possible, will evolve according to the projection context. For small indoor installations imperfectly dark rooms, the required parameters will not be very high, up to 2,000 lm, compared to more than 26,000 lm from several devices for 3D mapping on the façade of a building at night. Of course, to obtain images in which the viewer can immerse himself, other parameters are also important, like in cinema: the resolution, the aspect ratio or, sometimes, the weight and size of the projector.

Since projection mapping is generally much more than an effective presentation of visual layers and includes sound so that the whole thing creates an illusion of reality, it is essential that the sounds are properly designed. They include both non-diegetic musical compositions that define the character of the scene, and diegetic sounds that directly refer to the visions created (sounds of explosions, ruined buildings, burning roofs, etc.). If the sound is correctly synchronized with the image and is played on a properly adjusted audio system, the sound and image allow the spectators to fully immerse themselves in the vision presented to them, with the projections of fantastic worlds directly superimposed on the reality we know; worlds where diegetic, non-diegetic elements and mimetic spaces are mixed. A properly designed sound will make it possible to overcome the distractions of a street or even to transform space into a synesthetic, multisensory environment that penetrates the body and vibrates to the rhythm of sound.

The medium created following the design and projection of the projection mapping on an existing object known to the viewer (in the case of architecture) or a new installation also makes it possible to go beyond the image that is traditionally made of a screen. When a film is projected in a cinema, the screen, in a rather paradoxical way, disappears into the light of the projector. On the contrary, in the context of projection mapping, the screen – the object that has been mapped – will always be visible. Sometimes, and to a certain extent, animations will modify its shape to allow it to present its natural characteristics, and sometimes it will disappear more under visualizations that will totally modify its appearance. But this screen object will always be visible, its real and virtual 3D nature will be highlighted during the projection.

Thus, in the context of projection mapping, not only does the medium move from the flat, illusory surface of the cinema screen to a three-dimensional form that is always visible, but the space and context of perception also change. The work can be visualized in a separate, dark space, or in a public environment known from another perspective, with its natural implications and meanings and its surrounding natural sounds, for example in the case of urban architecture.

The animation design stage is much broader, less visible and less perceptible in the context of the projection mapping "dispositive". It involves obtaining the precise mask (technical, mathematical project) of the object on which the images will be projected, and then creating 2D and 3D animations synchronized with the sound and mounted as required to obtain the final shape that the spectators will see during the projection.

2.4. The spatialization of the gaze or the perception of the projection mapping spectator

As part of the study on the "dispositive" of projection mapping, viewers and the way they see and experience the audio-visual situation are very important. Spectators will always be suspended in a hybrid reality that brings together reality, existing 3D objects, mapped installations, architecture or any other element of natural public space, and computer-designed elements in the form of 2D or 3D animations that are projected onto real objects. This new context thus links real and augmented reality, and, as Baudry stated, creates an even greater "impression of reality" among the public (Baudry 1992). By using existing psychoanalytical tools to explain how people function, react, feel and experience their own position and reality, the French writer and essayist makes connections between film projection and a dream, adding that film projections are even reminiscent of dreams (Baudry 1992, p. 79). Referring to the myth of Plato's cave, he created one of the most representative metaphors of the psychoanalytical theory of cinema. Baudry nevertheless eliminated the parts on idealistic philosophical assumptions that material things are only the reflection of ideas. He used the cave metaphor to support his own theory, which he integrated into the psychoanalytical theory of cinema by affirming that external circumstances influence our state of mind.

According to Baudry, the prisoners of the cave, like the spectators in a cinema, see only the illusion of reality, the shadows of a ghost world. At the same time, the closed, secure and dark cinema room establishes a state of artificial regression that transports viewers to the prenatal period, into the mother's womb – a symbolic shelter that we are seeking to recover (Baudry 1992, p. 84). Projection mapping is not a direct extension of this psychoanalytical concept. Indeed, except for stage mapping, spectators are not bound by the dark cinema room or by seats that limit their movements and allow them to concentrate on the stimuli broadcast from the screen. Projection mapping spectators are suspended in between. On the one hand, the freedom of movement and even the possibility of walking around the space where the mapped screen object is located prevents a total commitment to the perception of the animation and the object. On the other hand, we refer to the "impression of reality" because the spectators' minds do not need to take them to the imagined places – they are already there or were even there before the projection began. The "impression of reality" is also integrated into cognitive abilities, including the human imagination. The technical images presented in projection mapping allow the spectators to go beyond the limits imposed by their imagination (often ignored during the dream) and to indulge in an unlimited fantasy. This corresponds to the fulfilment of a human desire: to have power over perceived reality and to co-create it, to modify it, without respecting the laws of physics. The viewer therefore experiences pleasure in looking at the images created, a feeling further reinforced by the fact that something unreal is added to reality, to existing objects. Vilém Flusser states that, thanks to technical images, we have at our disposal an almost unlimited power of imagination that "arises from our ability to calculate and measure the surrounding nothingness" (Flusser 1994, p. 59). This calculation will be particularly important in projection mapping, as it is based on mathematical calculations of scale, size, distance and general precision of animations, that is on mapping or projecting objects, copying, to achieve a modification effect during the projection.

During the perception of a large format video projection or smaller installations, the viewer's gaze is subjected to spatialization. Contemporary viewers are used to audio-visual projections, but in the form of films or animations projected on a 2D screen. 3D stereoscopic technology allows you to experience the depth of visualization modification with special glasses. It opened the possibility of a 3D perception of the images presented. Virtual reality technology further reinforces this impression of being in the reality of the film, which is similar to physical reality. This technology uses the spectator's entire visual field by separating it from reality using special glasses. Projection mapping offers a slightly different "spatialization". From the beginning, before the projection begins, the images are available as natural, physical, 3D objects (installations, scenography, architecture, etc.). But during the projection mapping process, the "spatialization" reorganizes itself on the mapped object, whose elements begin to change during the projection, to transform and to become an element of an additional narrative that disrupts its stable structure. The light of the projector, like the spectators, once again shapes the object observed with their vision, while they immerse themselves in the imagined reality.

According to the psychoanalytical concept of identification, the spatial screen becomes the double movement effect described by Metz: a projection movement – the people themselves project the image and are duplicates of the projector – and an introjection movement linked to the return of the image to the spectators' minds, that is their recording consciousness (Metz 1977).

A good example of the perception of projection mapping in the context of a large format projection on a public building is the projection mapping inaugurated in 2014 in Campeche, Mexico. The projections are made on the 80-meter façade of the library located in the city center. A half-hour show is presented every day, attracting both local residents and tourists. The projection mapping entitled *Celebremos Campeche* (Let's celebrate Campeche) projected on the building's biaxial arcaded facade is a story with animations presenting the history of Mexico, which begins with Mayan cosmology, other beliefs, colonization, until modern times with traditional beliefs (such as the cult of the "Santa Muerte") and religions brought by settlers (Christianity).

Figure 2.2. *Creative KAOS Management, Celebremos Campeche projection mapping on El Palacio Centro Cultural, Campeche, Mexico (source: photograph by Justyna Łabądź)*

The perception of this sound and light show is transformed when architecture becomes the screen. First of all, the spectators are in a public space they know. And even if they do not often use it, they associate it with integrated urban planning division rules of the city throughout the world, with horizontal roads, squares with typical "street furniture" such as streetlights, benches, flower boxes and vertical buildings that extend upwards and to the sides. They perceive this known architecture – now illuminated and modified by projection mapping – in two ways, from the point of view of their previous experiences, and in this new form that transforms its spatial division and allows them to introduce a new narrative, to reinterpret its facades, and sometimes to modify its character and the real history with light. Thus, paradoxically, it is thanks to architectural projection mapping that spectators can see the invisible screen in the cinema. But at the same time, with illusions, concrete or brick facades, walls and stable stones, seem to dissolve and dematerialize, like the stained-glass windows of churches. Projection mapping frees mapped objects from their structures, just as photography has freed painting from the mimetic reproduction of the world. It responds to our thirst for illusion and allows the realization of fantastic visions that have been found since the beginning of human history, so much so that man has always wanted to visually manipulate his reality, from the shadows of Plato's cave to modern cinema and augmented and virtual reality systems, as well as phantasmagoria and magic lanterns. Thanks to architectural projection mapping, not only the mapped object is reinterpreted, but also the space around it. Its "illumination", the enhancement of hidden elements, can even remind us of the magical atavistic practices that allow us to overcome the fear of the dark, the fascination for protective and benevolent light that eliminates the mystery that surrounds the surrounding space plunged into darkness. Cinema and its derivatives, including projection mapping, are nothing more than new expressions of the cult of the sun and light described in many cosmologies of the world, including the Bible, in the form of worldviews created through benefactor light. This light creates reality and makes it visible, while protecting us from the dark, the uncertain darkness, with its potential invisible dangers.

Another example of the involvement of spectators' perception are projection mappings, the progress of which depends on the actions of the audience. "RoomAlive", a project created by Microsoft Research in 2014 and unfortunately still in the prototype stage, is a very interesting example. "RoomAlive" is a game based on projection mapping technology that uses the entire room and its elements to create a visualization of augmented reality. Using autocalibration, he dynamically uses the room in which the player and his objects are located to place the characters of the game. Players fight against them kinetically, shoot in their direction, jump, advance or hit. In this type of projection mapping, the viewers do not only engage their sight and hearing. The immersion effect is reinforced by the fact that the entire body of the participant is involved, or "inside" the projection mapping. His involvement in the images he sees is therefore not only psychological, but also physical. As in other interactive audio-visual works, the participant becomes the co-creator and can decide how the animated story will unfold.

2.5. "Attractions set-up" or real content?

As can be seen from the examples above, projection mapping can take different forms and be used in different contexts, and this form of augmented reality that frees viewers from smartphone or tablet screens has a significant impact on the change in the overall layout of mapping projections and public perceptions. In the discourse of projection mapping, which extends and broadens the concept of augmented cinema, we will find new forms that radically modify the perception of images of techno culture, such as the experiments on holographic cinema mentioned by Youngblood (Youngblood 1970, pp. 399–414). Nevertheless, projection mapping creators still have a lot of work to do in order not to focus on the effects that often seem to dominate productions. Under the pressure of design, and within the framework of the domination of form over content, projection mapping represents the domination of the cult of space, with aesthetics and ideological discourses placed in the background, thus entering the critical vision of Baudrillard's simulacra. This approach, limited to the simple "assembly of attractions", considerably reduces the impact on viewers who are looking for real content that would allow them to remember the visualizations they have observed and their messages for a longer period of time.

Spectators will certainly not perceive the examples mentioned above in the same way, and these differences will depend on their own ability to perceive and get involved, the quality of the experience, and their previous experiences with similar projects. Each viewer will understand the images of techno culture in a different way, will choose the elements that interest them, that attract their attention, because often in the case of large format projections, they will be unable to perceive the whole project with all their senses (Dayan 1992, p. 87). We can use different methodologies to study them, but the most reliable would be cognitive methods: analysing the senses, the brain and the mind, using interdisciplinary knowledge from cognitive psychology (including the unconventional psychoanalysis methods mentioned above), neurobiology, mathematical models of the brain and brain philosophy. It is impossible to mention them all in such a short article. Nevertheless, this could inspire further discussions and texts examining the evolution of the way in which projection mapping is experienced and perceived, which, by its effects and form, undoubtedly remains very attractive.

2.6. References

Barthes, R. (1993). Wychodząc z kina, translated by Demby, Ł. In *Interpretacja dzieła filmowego*, Godzic, W. (ed.). Antologia przekładów, Krakow.

Baudrillard, J. (1983). *Les Stratégies fatales*. Grasset, Paris.

Baudry, J.-L. (1992). Projektor: metapsychologiczne wyjaśnienie wrażenia realności, translated by Helman, A. In *Panorama współczesnej myśli filmowej*, Helman, A. (ed.). Universitas, Krakow.

Dayan, D. (1992). Widz zaprogramowany. In *Panorama współczesnej myśli filmowej*, Helman, A. (ed.). Universitas, Krakow.

Deleuze, G. (1986). *Foucault*. Éditions de Minuit, Paris.

Flusser, V. (1994). Ku uniwersum obrazów technicznych. In *Po kinie? Audiowizualność w epoce przekaźników elektronicznych*, Gwóźdź, A. (ed.). Universitas, Krakow.

Gwóźdź, A. (2003). *Obrazy i rzeczy*. Film między mediami, Krakow.

Metz, C. (1977). *Le signifiant imaginaire, Psychanalyse et Cinéma*. C. Bourgois, Paris.

Youngblood, G. (1970). *Expanded Cinema*. E.P. Dutton, New York.

Webography

Room Alive, Scalable:

3

Projection Mapping: A New Symbolic Form?

3.1. Introduction

This chapter presents the dawn of research on projection mapping as a "symbolic form", based on the concept forged by Erwin Panofsky (Panofsky 1975), as well as the notion of "apparatus" developed by the contemporary philosopher Jean-Louis Déotte. From philosophy to aesthetics, from cinema to the fine arts, we will envisage projection mapping by considering its elements (surface, volume, projected images, projector) and its technical conditions to draw a landscape of its specificities.

3.1.1. *Symbolic form and apparatus*

The concept of "symbolic form" was forged by the German philosopher Ernst Cassirer. According to him, "the activity of the mind is a symbolic activity" (Rieber 2008, p. 1) within the field of culture. In 1927, art historian Erwin Panofsky approached the notion of symbolic form by applying it to a defined object. He was interested in perspective as a technical tool: starting from its geometric conception, he analyzed it as a founding element of a science and philosophy of space. From a mathematical representation system to the Cartesian model, perspective ultimately became a "mode of meaning for understanding the world" (Rieber 2008, p. 2). That is to say, an indication of the relationship existing between man and the world which is no longer interpreted as a religious key, as the manifestation of the power of

---

Chapter written by Martina STELLA.

God, but invented by man, and therefore "de-theologized". This presupposition would be developed, among others, by Daniel Arasse in *Histoire des peintures*: although the author denies Panofsky's definition of the symbolic form, he argues that "perspective builds an image of the world that is commensurable to man and measurable by man" (Arasse 2004, p. 65). Through the invention of perspective, "the men of time build a representation of the world open to their action and interests": through representation, man asserts himself as the subject who sees and acts. But what then do we mean by "symbolic form"?

The term "symbolic" comes from the Greek *symbolikos*, meaning "explaining with a sign". Following that there is a signifier, a form, and a signified, content: it is the form, and therefore the signifier will be endowed with symbols that help to nuance the signified and charge it with meaning. Perspective being a means of "seeing through" or "seeing clearly" (Panofsky 1975, p. 37), it is an objective and accurate tool for representing space.

For Panofsky, the conception and perception of space are constantly evolving human apprehensions: he thus establishes links between the evolution of human conceptions and the evolution of the representative system. The invention of perspective, the first system of representation, is thus symptomatic of a collective state of mind. In other words, each system of representation reflects the emergence of a new collective mentality and consciousness – of a new belief, perhaps? Changes in the relationship between the subject and the world, surely.

This is how the discourse of the contemporary philosopher Jean-Louis Déotte intervenes, enriching the reflection on the symbolic form with the notion of "apparatus". Perspective is the first "projective apparatus", in a succession of "technical apparatus of modernity such as perspective, *camera obscura*, museum, photography, cinema, psychoanalytic cure, etc., which initially constitute the conditions of the arts, period after period" (Déotte 2008a, blurb). Each apparatus can be conceived as a symptom of a social state, but also of a temporal index being intrinsically linked to it.

The definition of camera from which Déotte formulates his theory is the one advanced by Vilém Flusser, in his theoretical work *Pour une philosophie de la photographie*:

> "The Latin word *apparatus* comes from the verb *apparare* which means 'to prepare'. Latin also includes the verb

praeparare, which also means 'to prepare'. If we want to capture the difference between the prefixes *ad* and *prae*, perhaps we could translate *apparare* as 'to get ready'. From then on, an apparatus would be a thing held ready that is on the lookout for something. [...] An attempt to etymologically define the concept of apparatus allows us to point at the 'being-ready' attribute specific to the apparatus, as at the rapacity that characterizes it." (Flusser 1996)

According to Jean-Louis Déotte, when "apparatus are the technical and cognitive conditions of the arts" (Déotte 2004, p. 50), the apparatus is an indication and catalyst of the "being ready", a historical and cognitive change. The apparatus determines both the "appearance" of a temporality of its own and an *apparatus*, that is, it configures an "organized set of systems" because "the apparatus is what articulates the sensitive and the law in the form of an address to the singularity and to the being in common" (Déotte 2007, p. 23).

The apparatus is a technical tool that lays down the conditions of the arts and aesthetics, since "any technique that develops an autonomous game on perception and sensitivity, [...] affecting a singularity, transforms it" (editorial committee, *apparatus*, presentation). By acting on the sensitivity of the individual as well as on the state of the mass, by referring to the surface of inscription and writing of the law of an era, the apparatus also reveals itself as a political tool.

3.1.2. *Apparatus and projection mapping*

In the light of these considerations, we would then like to put forward a new analytical hypothesis: can we insert projection mapping into the continuation of these projective apparatus, the most recent of which are the museum, the video and the artistic installation? (Déotte 2018, p. 17). One of the main reasons for the interest in this particular hypothesis lies in the notions of *continuation* and *continuity* between the different apparatus because "it is not a single device that makes a difference, but a plurality or a conjunction of apparatus" (Brossat 2012, p. 7). We will approach projection mapping throughout the notion of *apparatus*: more precisely, we will consider projection mapping in its interactions with the apparatus studied by Jean-Louis Déotte, in order to better understand its specific characteristics.

3.2. A shifting tool

Projection mapping is primarily a technical device, a tool that enables projection on volumes. The verb *to map* recalls the act of drawing a map, as well as the mathematical "application", a function associating two elements.

In projection mapping, the two meanings function almost as synonyms: the act of mapping refers to the projected image that is mapped, drawn, appears, adhered to, *applied* to the surface of an object.

Mapping consists of an *action*, carried out using a projector and software, that creates correspondences between a video and the shape of an object. The technical characteristics of the projector, as well as the choice of software, are decisive in the realization of the device.

The projected video can be made according to the medium, from the template of a building for example, in the form of an animation video. On the other hand, considering projection mapping in terms of its technical nature, means that the video content doesn't need to be created according to a specific material. The principle of the tool allows projection mapping to locate any type of *moving-image* on an object. As long as the software supports video file formats, we can process and map archive images, mainstream films, or experimental sequences on media. Certainly, the lack of dialogue between form and content, and therefore between medium and video, is detrimental to the power of the installation. But it is no less important to note that the principle of the device also allows us to adapt to the projection of video content that has not been created specifically according to the mapping.

The object, the medium of the projection, can be the model of a building or the building itself: the scale as well as the shape of the medium are variable. The common and fundamental characteristic of all projection mapping installations is the need for a projection space protected from direct sunlight, a problem linked to the luminous power of the projectors; the other projection constraints being specific to the installation site.

Projection mapping is a metamorphic and transversal practice. At the crossroads of art, technology and communication, it is a multifaceted and versatile technical device: its status and meaning depend on the environment in which it is projected. A shifting tool: like these linguistic categories that introduce a transmitter and a receiver into the enunciation, but whose meaning is only revealed in a specific oral context. Projection mapping is a versatile projection technique, which has no form in itself, but is built from the context, therefore the medium, on which the projection is made.

The notion of *medium* concerns the characteristics of the object hosting the projection, its appearance, the materials that compose it, its presence in space: it can be a fixed medium or a mobile medium, animated or inanimate, visible only when illuminated by the projection, or perceptible independently of it.

However, the medium also refers to the context in which the projection occurs: the form that the projection assumes takes on a different meaning depending on the environment in which the projection takes place. To speak of the heterogeneity of the circumstances in which this engaging tool participates is to consider this practice in theaters, as digital scenography; in exhibition halls, as much as a work of art as an instrument at the service of managers; in amusement parks, in events, in the cultural industry in the broad sense, but also in didactics, in education, in data visualization, particularly in geography, architecture and urban planning. Outdoors, in public spaces or indoors, spectacular character is preferred over the use of other practices: the meaning of projection mapping must be nuanced according to the context in which it is presented, which, moreover, gives it a specific function and a proper status.

As the shift consists of a transmitter and a receiver (for example, the demonstrative *that*, the pronoun *I* or the adverb *here*), the mapping is based on the image/medium dichotomy of the projection. The surface of the medium is emblematic of the in-between, as a moment/space where two heterogeneous realities meet. We can speak of the co-presence of two temporalities, where the duration of the projection meets the temporality specific to the object: in an ephemeral way, the projected image inhabits and covers the surface of the medium. It does not consist purely of an amplification of the nature of the object, but of a palpable temporary transformation of its essence, like a disguise. By masking certain parts of the volume, the very shape and qualities of the medium object can change. It is, of course, a temporary transformation, like the chameleon that changes the color of its skin according to the object on which it rests. But unlike this animal, whose goal is to protect itself by camouflaging itself in its surroundings, the relationship between mapping, vision and the environment is not a matter of mimicry: it is rather a matter of appearance, disguise and display.

Projection mapping is therefore a technical and aesthetic tool that stimulates a certain sensitivity, working like a shifter. A system whose meaning depends on the context and the conditions of visibility it establishes. We will now focus on the two elements of the dichotomy that form the basis of a projection mapping installation: the surface that hosts the projection and the projected image.

3.3. The surface

3.3.1. *The environment/projection ratio*

One of the characteristics that differentiates projection mapping from a simple projection is that the software specific to this technique allows images to correspond to surfaces in space on extremely variable scales.

These softwares are of fundamental importance in the definition of projection mapping; a specific study should be dedicated to the different devices that have been developed since the 1930s, as well as the types of software that are used today. Projection mapping is currently both a physical device and a digital software. However, we will focus on the notion of device, to analyze the relationships between the image and the medium through the projector.

Since the end of the 19th Century, images are spreading, increasingly inhabiting the urban environment: cinema, the artistic world and advertising have contributed, on different scales, to the emergence of their presence in public space. In the case of cinema, for example, the first screening in a public place dates back to December 28, 1895 when 10 of the Lumière brothers' films were screened at the Indian Salon du Grand Café in Paris. The first public screening of a film takes place only a few months after the invention of cinema. The same applies to advertising: billboard advertising was born at the end of the 19th Century and gained momentum at the end of the 1950s (Martin 2004, p. 59). An increase in the size and number of images in the city can be noticed: they have become omnipresent and have now been normalized in the eyes of its inhabitants. Almost simultaneously, in the 1960s, urban art appeared, a movement that took the work out of the gallery and integrated it into the day-to-day environment.

Little by little, the image thus emerges from the verticality of the wall, its frame, its status as a representation, to be part of the environment. At the foundations of projection mapping too, we can discern a series of devices and experiments for spatialization of the image. The desire that moves many practitioners working with images is also reflected in the theatrical environment, where one of the first devices of spatialization of the image was realized: it concerns the placement of screens in space. In 1958, the Czech scenographer, Joseph Svoboda, presented the installation of the *Polyécran* at the Brussels exhibition. In a dark space, many screens guided the spectator in the dark. Each screen had a corresponding projector: there were no actors in this immersive installation: the screens were the landmarks and guides for those who walked through space.

The screen, to use Éric Bonnet's expression, has become a "place of images" (Bonnet 2013). The screen has finally become a place in its own right. Is it a phenomenon of ubiquity, between one's own physical presence and one's virtual presence? From this perspective, projection mapping advances a silent invitation to bypass the screen. By this technique, the spectator no longer needs a digital interface: the projection unfolds on a volume that is there before it adheres to it and that will remain there once the projection is finished. If before we could have believed that the screen was becoming a space, a place, now we can see the reversal of this trend: the space becomes a screen.

3.3.2. *The volume*

By projecting onto volumes, we are quickly moving away from contemplating the model of the rectangular, flat and white film screen. The boundary of a screen is its frame, whether it is a television screen, one or more computer screens, the surface or fabric on which the film is projected: the frame functions as the frame of the image, both supporting and limiting it.

Daniel Arasse defines the framework as "what from what, and not what through what, one can contemplate" (Arasse 2004, p. 97). Referring to antiquity, he explains the root of the term contemplation by the Roman *templum*: a rectangle in the sky inside which the oracles read the flight of the eagles.

As in the cinema, the frame of a projection mapping projection depends on the area that the projector or the arrangement of several projectors can cover. Since film projection is made using the same tool, the specificity of projection mapping lies in the mediation of software, which allows images to be precisely located in space. Among the fundamental operations that projection mapping software updates, we must mention "masking", which makes it possible to "select" the portion of space on which the projection acts, as well as *soft edge* or *stacking*, arrangement of projectors to arrange several of them to obtain a monumental image, the size of a building. The notion of a frame is lost in the real dimension of the object it receives.

Even the most recent appropriations of the screen concept, such as Philippe Parreno's projections of Rauschenberg's[1] *White Paintings*, as well as Michael Snow's[2] *Vue euv*, are overwhelmed by the world of projection mapping where any object is a possible screen. It is no longer necessarily monochrome, nor necessarily flat: it only needs to be material and sufficiently opaque.

Projection mapping, unlike other digital devices, requires a medium other than the screen interface to exist. All the work prior to the projection is done according to the object and the context in which the projection is expressed. The creation of video content, positioning and calibration of projectors will be all the more effective if they are reflected as a function of an instant t in a space (x, y, z). The object on which the projection is expressed is decisive in the meaning and reception of the video. A monumental projection, for example, makes a building – and therefore the architecture of the city itself – the object to look at.

Through projection, almost any object can become a potential medium/surface. Of course, design and software require prior work, but during projection all digital work makes sense only in the encounter with volume and medium.

---

[1] Installation presented in 2002 at the Musée d'Art moderne de la ville de Paris, on the occasion of the Alien Seasons exhibition. The work consists of the seven panels composing Rauschenberg's *White paintings* (1951), on which P. Parreno's film *El Sueno de una Cosa* film was projected; between one projection and the other, *4'33"* of silence recalled John Cage's work.

[2] Michael Snow (1998). *VUEEUV*, photographic printing on fabric, 103 x 130 cm.

It is interesting here to mention the question of *immateriality* raised by Jean-François Lyotard concerning the advent of digital technology. In the exhibition entitled *Les Immatériaux*, organized in 1985 at the Centre Pompidou, the staging leads the spectator to question the evolution of man's relationship with the world – at a time when the words "contemporary" and "technical" began to be consubstantial. The notion of "immaterial" refers to the digital domain and the digitization of data. We cannot clearly distinguish the shape of the material here, because the machine is both the place of creation and the place where information is disseminated: the binary language translates the image, sound, text or video, discreetly. The boundary that previously coincided with the screen, the interface between virtual and actual, is dissolving. If the idea of a frame no longer seems particularly relevant, thinking of the idea of a contour as an alternative may be valid. It is Gilles Deleuze who, in the *Logique de la sensation*, writes: "What separates and unites both form and substance is the outline as their common limit" (Deleuze 2002, p. 115).

3.3.3. *The projection plane: the substrate*

Reflecting on the notion of medium in mapping, it is no longer a question of thinking about the plane of the canvas as in painting, nor about the projection screen as in cinema: the notion of screen as an interface is missing, the projection is no longer delimited by a standardized frame, but by the outlines of real objects. As images dress up space and things, the relationship to the inscription medium becomes by metonymy the relationship to reality.

We could thus speak of the inscription surface as a contact plane, or even an opening plane between reality and projection, surface and image, material and immaterial, tangible and visible. According to Jean-Louis Déotte, the concept of "inscription surface" as a definition of the painting was overthrown by the advent of impressionism. His discourse is based on the comparison between the relationship that the perspective construction and a painting by Cézanne establish with the *plan of the painting*. Through perspective, "the painting plane is a plane of projection" (Déotte 2007, p. 26), which opens up another dimension by becoming transparent. Painting brings us to its interior, to the space built according to the laws of the system of perspective representation. Alberti's famous *"window"* creates an alternative by making us forget the two-dimensional dimension of the canvas. In contrast, in a painting by Cézanne, the plan of the painting functions as "a membrane, which allows reciprocal exchanges between the *subject* and the *object*" (Déotte 2007, p. 26).

In this regard, it is very interesting to quote a letter from Antonin Artaud to André Rolland de Renéville dated September 2^3, 1932: "Here included is a bad drawing in which the so-called *subjectile* betrayed me" (Derrida, Thevenin 1986, p. 55). However, Artaud reverses the order between surface and writing to show that it is his drawing that does not satisfy him, certainly not the shaping of the sheet on which he wrote. The notion of the subjectile, taken up by Jacques Derrida, has a profound and quite specific interest because it refers both to the subject, to the referent in the Barthesian sense, and to the medium of representation. If we apply this perspective to projection mapping, we can notice that the same *double inclusion* relationship is created between video and medium (Brossat 2012, p. 3): projection mapping could be considered as a subject-matter. It is neither the projection nor the medium itself, but the in-between.

3.4. The projection

3.4.1. *The haptic image*

As screens dissolve, they no longer constitute the interface between digital and the environment: projection makes it possible to create direct contact between intangible and tangible data. Paradoxically, it is not the hand that touches the objects, but the image, the representation itself, which acquires this double haptic sensitivity by the way it meets the surface. The projected digital image becomes a (naturally ephemeral) presence, because the time of experience is the time of projection, but directly related to the medium in its materiality.

The sensation of a "tactile-optical space" described by Deleuze is created by the encounter of two realities, two temporalities, two materialities: "It is not simply visual, but refers to tactile values, while subordinating them to sight" (Deleuze 2002, p. 118). In the situation created by mapping, one could imagine that it is no longer simply the eye that acquires this ability to touch the image, but that the image itself, in the form of light beams, touches the

---

3 A "subjectile" is an "external surface on which the painter applies a layer of plaster, paint or varnish".

object on which we project it: the representation maps the medium as a layer of light that touches the surface, giving a tactile dimension to the image.

3.4.2. *The point of view or the projector*

From the image to its source, we will now focus on the projector and its role. In the realization of a projection mapping installation, the image is calculated according to the place of the projectors, which function as points of view from which the space is filled with light. The point of view is a concept that comes from the perspective construction, in which it coincides with the position of the subject looking, it *is* the observer: it embodies the position (physical as well as metaphorical) of the one who looks (Damisch 1987). In a projection mapping installation, the projector is found in this position, as a single-eye device from which the image is projected. A projection can be updated from the "points of view" of several projectors: by *stacking* and *soft-edge* techniques. The projector, as a source of a reproducible image and as a tool generating a point of view from which the image is constructed, can be interpreted as a symbolic tool, as an emblem of the fragmentation of perspectives that characterizes the post-modern era. Svoboda was already talking about the *Polyécran* when he said that it was a device "in the image of our complex world devoted to the 'bombardment' of simultaneous information" (Bablet 2004, p. 128).

In cinema also, the projector coincides with the point of view from which the film takes place; the vision, for the spectators, is shared by involving "the overthrow of the simple spectator as a practitioner of the collective" (Déotte 2008a, p. 9). Projection mapping and cinema do indeed share many characteristics, "in their relationship to the natural environment (visibility of movement, dilation of time) and in their relationship to the human environment (the political aspect of cinema, the overthrow of the simple spectator as a practitioner of the collective" (Déotte 2008a, p. 9). But unlike a film projection, which is always frontal, projection mapping works by aligning the projector's point of view with the object, concerned by the principles of parallax and anamorphosis. The anamorphosis is an optical phenomenon of image distortion, produced by means of a curved mirror. This phenomenon of distortion has often been represented, particularly in painting: one of the best-known examples is the painting *The Ambassadors* by Holbein, dating from 1533. Contemporary artists Felice Varini and Georges Rousse use this technique, for example, in order to create *in situ* installations. From the interaction between the spectator's position and the painted shapes, the alignment of the perspective is updated and the image appears seen from a certain angle. The adhesion of the image to the medium, within a projection mapping installation, acts according to the same protocol by working with layers of video rather than flat colors.

The projector establishes the point of view from which the image is created. We can discern this concept in it the notion of *"apparatus* that configures the common sensitivity": within a projection mapping installation, the projector configures, indeed, the spectator's way of looking. The spectators find themselves forced by anamorphosis to adopt a certain point of view, but at the same time are nomads, free to circulate in space and create their own points of view. The free wandering of the spectators during the screening is another characteristic that differentiates projection mapping from cinema. Déotte reminds us of the distinctions made by Anton Ehrenzweig in *The Hidden Order of Art* (Ehrenzweig 1982, pp. 39–55) between a conscious and focused, and therefore *active,* attention and a more erratic, "sweeping" look: therefore rather passive or "reactive" to the surface and its hidden layer. This erratic gaze, *scanning*, or even nomadic gaze, is a reaction to the many visual inputs that Déotte describes as "uncontrolled visual displacement [...] the only one that can bring up colored or material events. A perception of texture". Projection mapping responds well to this conception of a look that perceives texture, in the way that it interacts and relates to the inscription surface.

The relationship between sensitivity and time, which Déotte highlights, consists of a perpetual exchange between the apparatus which determines a certain "configuration of mass" and the conditions that mass itself allows the apparatus to update: the apparatus springs from a society, whose sensitivity will be built from the apparatus at its disposal. But since the apparatus "addresses the singularity and the being in common" (Déotte 2007, p. 23), it is also necessary, precisely, to consider the relationship it maintains with the community. From this point of view, projection mapping invites us to take a collective look at a visual and ephemeral transformation of reality.

The specific feature of this tool is that, unlike virtual reality, projection mapping's projection works, with few exceptions, without glasses that isolate the spectator, but in a space that invites us to share the experience.

Projection mapping confronts the spectator with two visual elements, the image and the medium, each loaded with its own meaning: from the dialogue or rupture between the shape of the object and the content of the video comes the subject-matter. The object on which the projection is expressed is thus decisive in the meaning and reception of the video. In the case of a monumental projection, for example, it is a building, but also the city which, by metonymy, becomes the object to contemplate. To attend a mapping projection is to look at the object on which the light intervenes. We are thus led to look through a one-way mirror, with the light coming from the outside and its reflections appearing on the surface, while this light we can see the inside of the brightened room emerge in transparency.

3.5. Conclusion

Projection mapping thus updates a phenomenon of emerging, an action of transformation, ephemeral and superficial, of the essence of this object. The projection intervenes on a still surface and animates it, arousing a feeling of surprise and astonishment.

The reaction to these installations can be compared to the "wow effect", the effect of surprise, admiration or appreciation following the discovery of a product in the marketing world (Wow effect, marketing definitions). It should be recalled that the first real application of this technique took place in a Disney amusement park within the *Haunted Mansion* facility in 1969. Disney can be considered, from the late 1960s until the appearance of augmented reality in 1998, as the pioneer of technological research in the field of moving image projection. The technology behind video mapping has therefore been developed by the cultural industry community: first in amusement parks, then in light festivals, but also, as we have seen, in theatrical scenography.

The projection mapping technique is gradually escaping from these fields to other uses, becoming a didactic tool, a medium contributing to the mediation and preservation of cultural heritage, an instrument at the service of collective memory, or a new form of expression in the contemporary art world. The question is whether the relationship to the "spectacular" is one of the fundamental characteristics of projection mapping or whether, through its propagation and adherence to other environments, a new relationship between projection and the spectator will be established. On the one hand, its diffusion could lead to the standardization of this practice: like the advertising images on the city walls that we have been able to talk about before, the spread of the phenomenon prevails, from omnipresence to trivialization. At the same time, as a technological tool in the service of didactics or science, its use would prevail over the message it contains rather than over its own form. In data visualization, for example, in the service of geography, urban planning or architecture, projection is no longer a tool for demonstrations that arouses wonder, but a visualization of the values expressed by the data themselves. In this sense, the development of projection mapping in other environments, particularly in the scientific field, could change the attributes intrinsically linked to the environment from which the projection mapping comes, particularly its spectacular nature.

In this regard, it is interesting to mention the research developed at the MIT Media Lab Studio by Yan Liu, Yihang Sui and Jingxian Zhang under the direction of Kent Larson. The project, called *Model Cities*, uses projection mapping to make the megadata of an urban site visible (from English: *Urban Big Data Visulatization and Projection Mapping*). The purpose of this research is to study the quality of public spaces and the level of urban safety. After collecting the data, they worked on a 3D model built with Lego to project the parameters from the *in situ* and *online* surveys, whose values were translated into color using a legend. This project refers us to the research of the English landscape architect James Corner, as well as Nadia Amoroso, specialists in the visual representation of urban phenomena. We are interested in these two personalities because they are developing a new concept of the mapping act, which aims precisely at visualizing data through the illustration of maps. These two researchers do not work with projection mapping, but rather focus on 3D modeling and drawing. However, the desire that drives them is very similar to the one that inspired Yan Liu in his project at the Massachusetts Institute of Technology (MIT): mapping to integrate and represent intangible data in space. For Yan Liu, projection mapping is an instrument at the service of urban planning, architecture and sociological surveys, making it possible to make intangible data visible, by making them visible and intelligible.

Another emblematic example is the project of the Chamber of Trades and Crafts of the Tarn-et-Garonne department, which has been working on the *Artisan numérique* since 2001. This research project has given rise to the *Répertoire numérique du geste artisanal*, which aims to revive – or even survive – the crafts and gestures of artisanal creation through digital tools, particularly projection mapping. After filming craftsmen at work in their workshops, videos of the gestures and steps of manual work are projected on the city walls[4].

In this context, projection mapping is a tool at the service of collective memory, for the preservation of intangible heritage. The choice to celebrate the craft, intangible heritage, using digital tools, is a symbolic gesture. Thanks to projection mapping, the know-how of a gesture in danger of being forgotten is made visible, material to the eye. Projection mapping brings the spectators together around the projections in the city, creating in the public space an educational reality, a celebration of memory, a manifestation of resistance to the passage of time. The public space becomes a gathering place *for* and *through* images. And as images become an event, the city becomes a place of images (Virilio 1988, p. 47).

Projection mapping acts as a political tool, bringing the community together in public spaces and creating a new sensitivity, a new relationship with the environment and the eye. As we have seen, projection mapping is an engaging tool that can be found in quite heterogeneous situations. Each installation is based on a surface-to-projection ratio specific to oneself, creating meaning through its subject matter.

This projection technique is therefore characterized by a new relationship to the inscription surface, which is hidden by the projection while being revealed by light. The encounter between the medium and the projection also establishes the co-presence of two heterogeneous temporalities, hence the idea of a synchronic time that unfolds during the projection. The duration of the projection also determines the emergence of a haptic image, a tactile and optical image, which inhabits the surface of the object by temporarily transforming its visual nature. This action of transformation, this action of molding accomplished by the light layer, creates a wow effect that brings the result of the projection closer to a phenomenon of the order of the spectacular, while enabling us to see the underlying object differently. That is to say, during the projection, we are looking at the images-movements, of course, but also, implicitly, at the latent object that gives them a form.

As Déotte writes, "any technique that develops an autonomous game on perception and sensitivity, [...] affecting a singularity, transforms it" (Déotte

---

4 See, in this regard, the exhibition *La croisée des gestes* which took place at La Cheminée, in Septfonds, in 2017.

2008b). Once the projection is finished, the object remains. As we were able to evoke in support of the exhibition conceived by Jean-François Lyotard in the 1980s, the questioning of immateriality was part of a concern for the loss of the direct relationship to things: the relationship to matter, to tangible objects. From the detachment of the artisanal practice to a relationship with objects mediated by the technical tool. Subjectile mapping, the art of the in-between, creates a collaboration between the materiality of our environment and the immateriality of digital means. While projection mapping can be considered as a symbolic gesture, it is perhaps also for the relationship between the digital and the tangible that it establishes.

3.6. References

Arasse, D. (2004). *Histoire des peintures*. Denoël, Paris.

Bablet, D. (2004). *Joseph Svoboda*. L'Âge d'Homme, Lausanne.

Benjamin, W. (ed.) (1989). L'œuvre d'art à l'époque de sa reproductibilité technique. In *Écrits français*. Gallimard, Paris.

Bonnet, E. (2013). *Esthétiques de l'écran, lieux de l'image*. L'Harmattan, Paris.

Brossat, A. (2012). *Entretien avec Jean-Louis Déotte. Jean-Louis Déotte, Walter Benjamin et la forme plastique. Architecture, technique, lieux*. L'Harmattan, Paris.

Cassirer, E. (1920). *Théorie des formes symboliques*. Bruno & Paul Cassirer, Berlin.

Couchot, E., Hillaire, N. (2003). *L'art numérique. Comment la technologie vient au monde de l'art*. Flammarion, Paris.

Damisch, H. (1987). *L'origine de la perspective*. Flammarion, Paris.

Deleuze, G. (2002). *Francis Bacon, logique de la sensation*. Le Seuil, Paris.

Déotte, J.-L. (2004). *L'époque des appareils*. Lignes & Manifestes, Paris.

Déotte, J.-L. (2005). *Appareils et formes de la sensibilité*. L'Harmattan, Paris.

Déotte, J.-L. (2007). *Qu'est-ce qu'un appareil? Benjamin, Lyotard, Rancière*. L'Harmattan, Paris.

Déotte, J.-L. (2008a). *Le milieu des appareils*. L'Harmattan, Paris.

Déotte, J.-L. (2008b). Le musée, un appareil universel. *Appareil* [Online]. Available at: https://journals.openedition.org/appareil/302?lang=en.

Déotte, J.-L. (2012). Les immatériaux de Lyotard (1985): un programme figural. *Appareil* [Online]. Available at: https://journals.openedition.org/appareil/797.

Déotte, J.-L. (2016). Benjamin et le paradoxe de la chaussette. *Appareil* [Online]. Available at: http://journals.openedition.org/appareil/2354.

Déotte, J.-L. (ed.) (2018). Simondon: les appareils esthétiques entre la technique et la religion. In *Cosmétiques: Simondon, Panofsky, Lyotard*. Des maisons des sciences de l'Homme associées, Plaine Saint-Denis [Online]. Available at: https://books.openedition.org/emsha/222.

Déotte, J.-L., Ramond, C. (2008). À propos d'Appareil. *Appareil* [Online]. Available at: https://journals.openedition.org/appareil/.

Derrida, J. (1986). Forcener le subjectile. In *Artaud, dessins et portraits*, Derrida, J., Thévenin, P. (eds). Gallimard, Paris.

Ehrenzweig, A. (1982). *L'ordre caché de l'art*. Gallimard, Paris.

Flusser, V. (1996). *Pour une philosophie de la photographie*. Circé, Paris.

Lyotard, J.-F. (1985). *Discours, Figure*. Klincksieck, Paris.

Martin, M. (2004). De l'affiche à l'affichage (1860–1980). Sur une spécificité de la publicité française. *Le Temps des médias*, 2, 59–74 [Online]. Available at: www.cairn.info/load_pdf.php?ID_ARTICLE=TDM_002_0059.

Panofsky, E. (1975). *La perspective comme forme symbolique*. Editions de Minuit, Paris.

Rieber, A. (2008). Le concept de forme symbolique dans l'iconologie d'E. Panofsky. Reprise et déplacement d'un concept cassirérien. *Appareil* [Online]. Available at: http://journals.openedition.org/appareil/436#tocto1n1.

Virilio, P. (1988). La lumière indirecte. *Communications*, 48, 45–52 [Online]. Available at: www.persee.fr/doc/comm_0588-8018_1988_num_48_1_1719.

Webography

Urban Big Data Visualization and Projection Mapping:

4

Points of View: Origins, History and Limits of Projection Mapping

4.1. The origins of a movement towards alternative forms according to Romain Tardy

4.1.1. *Origins and VJing*

Romain Tardy participated as an author in the emergence of a trend that took the name of projection mapping. First, under a collective signature with Anti-VJ, then in his own name, before making other works of light that he says are not projection mapping. When Romain Tardy first mapped in 2006, it is not certain whether the term "video mapping" was used. This took place in a hangar where projections on crosses were made. In 2007 and 2008, on the other hand, the word *mapping* was used. It is difficult to remember his influences at the time, he says, but it was certainly very exciting to experiment with this form; everything seemed new, although with Peper's ghost, there was already the idea of increasing a reality and Tony Oursler or Michael Naimark in the 1990s were using digital to do projection mapping.

While projection mapping comes from VJing (manipulation of digital imagery in real-time) for some, it is a question of generation. And, notes Romain Tardy, without himself knowing what this is due to, a community has formed according to the VJing. This network database of electronic festivals helped to develop the mapping. Machines were quite expensive at first for an independent artist, he also recounts, but then access to equipment changed the situation with the democratization of technological tools.

Chapter written by Ludovic BURCZYKOWSKI and Marine THÉBAULT.

He learned from this period that VJing is very fragmentary and that it is a particular way of thinking. Romain Tardy created video clips assembled live, and there was no narrative or film logic. He pointed out that it is liberating to experiment with very short formats: you can experiment very quickly. Romain Tardy now finds the process more interesting than the result at the time and admits he is not very proud of everything he did. But he found limits to this fragmented process and therefore stopped it to achieve something more linear. Moreover, mapping today is more often built like a film, and VJing, he thinks, is probably less interesting artistically from the point of view of a spectator. The VJing shows an acceleration of life with its fragmentation, whether it is attention or even encounters. And, in doing so, he also does what he sometimes denounces. Saturation of attention, for example, is sometimes assumed, and at other times, it is in contradiction with what is denounced in this environment. Romain thinks that other fields should do projection mapping but do not do so because of technical familiarities in particular, and find themselves blocked to take the step when they have ideas and would do good things in his opinion.

4.1.2. *Transformation and continuity*

Romain Tardy studied at the Beaux-Arts with a visual artist approach. He was already doing installations with objects, even if he wasn't necessarily doing video projection in space. He then discovered the computer and did 2D animation. Then, he switched to VJing, and at the beginning of the projection mapping, with the anti-VJ collective, he made sound and visual installations called "multimedia", even if he doesn't like the name. It was about human performance.

The artist who does projection mapping is first and foremost an artist, says Romain Tardy, and the term "mapper" to describe the person who does projection mapping may seem a bit old-fashioned. Romain Tardy does not call himself a *mapping artist* because he does less now and there is the problem of referring to what he did in the past without leaving any opening in the future. He prefers to say *visual artist* which he finds better than *digital artist*. In French, he says, he has the impression that there is a division between "digital" and "visual" and that he notices less of this difference in English when using *visual artist*.

Romain Tardy therefore makes less projection today; he focuses more on structures that emit light and not only on mediums that receive it. "We don't often show the source of light with projection," he says. Now it reverses: it is the object that emits. Enlightening people was part of the things he did before and "when a context comes to light, I like it," he says. Regarding the common points between projection mapping and the work he is doing today that are not projection mapping, Romain argues that the context is still strong. The place is the theatre of the work, he says, it is formally dependent on it. In projection mapping, it is technically impossible to escape the support. Now Romain Tardy is in transition between *in situ* and autonomous object. When a structure planned for a site is moved elsewhere, it makes the original site travel. It wasn't thought of at first, "but when I think about it now, I like it," he says. The work can live *a posteriori* from its initial context: it is not in an exclusive relationship with a place. Romain Tardy does not forget the initial context, but does not refer to it continuously and it is a sufficient level of abstraction that allows him to do so, like a ghost allegory. He doesn't quote directly from architecture, for example, he doesn't find it interesting. But just a little reference to the initial context is poetic and beautiful. It is necessary to find a balance between something that is neither too invasive nor too "humdrum" in reference to the context. There are other challenges today that drive him, even though he admits that he may have felt freer before: he was experimenting.

4.1.3. *Projection mapping and the screen*

Romain Tardy has produced work that can be presented either in projection mapping or on a more traditional screen. "We have been used to the page or board for centuries," he says; the screen is a cultural habit more than an object. For narrative purposes, the screen remains unbeatable but there is an additional sensory layer in the installations, thanks to the fact of being able to stand or to use all five senses. He wondered to what extent one could have both an intellectual approach and a feeling. Body relaxation facilitates certain possibilities but the screen is more intellectual, cerebral: it requires more attention, or effort, perhaps. It is an immersion that comes from within, while immersion through context comes from outside and is more striking.

As for ephemeral projects, he is doing less now, and is making sure at least that he has the right conditions to do so. The mapping that is done and seen only once, in *one shot*, is the most difficult in his opinion. This does not exist in other arts: a performance can turn. This is an extreme situation: there is no second chance and only some document remains. It's a real particularity, the unique diffusion, it's necessarily frustrating and exhausting, untenable in a career. "As soon as you are demanding, you can of course always repeat your work," he says, "and in this case it is impossible to improve. That being said, as soon as it happened, it happened. And if it wasn't so good, it'll go away and we were able to try something. Leaving the screen also applies to the time spent in front of the screen," says Romain Tardy: "I'm not a big fan of studio work, I think it's better to be in the field." He prefers to be more in "reality" than in "increased reality".

4.1.4. *Projection mapping of yesterday, today and tomorrow*

Romain Tardy finds it unfortunate that mapping artists today can spend little time on site. In the beginning, *in situ* work was very important, probably more so than today in the sense that tools allowed less and on-site preparation was therefore inevitable. Working more remotely today is possible, as it has been optimized with plans and simulations. But before, it was necessary to spend nights on site, with a feeling of closeness to the more intimate place. The evolution of mapping is also perhaps less spectacular from one year to the next now; it seems more difficult to be original, thinks Romain Tardy. Clearing requires differentiation and it is less easy to have the effect of breaking away from the impression you had before. However, he does not think that the issue is a technical challenge now, but that it is more a question of looking at what art is in the public space; Romain Tardy asks how to overcome the harmless spectacle, how to put his art at the service of political issues, the environment and society. Without dwelling too much on the "society of the spectacle", if everything is reduced to entertainment, where are the spaces to talk about what concerns us? Transforming public space with projection mapping is something "public", so has a huge potential to open up public debate beyond the artistic field, even if it is art that produces and causes gatherings. What Romain Tardy finds exciting in our time is the decompartmentalization of disciplines, institutions and places. And it is the responsibility of the authors and organizers to ask themselves questions about the rallies and to ask themselves how far they can go.

As for the future, it is bleak, according to Romain Tardy. But it is not pessimistic: collaborative processes are developing. Projection mapping has become industrialized and it is often in the hands of the same people, who are less inclined to experiment than others. This industrial approach was nurtured by the pioneers of which he was a part, that's for sure, and he recognizes that you can't always experiment. However, he argues, it is necessary to find a balance between responding to orders and continuing experimentation. In art school, he opens, it is too little addressed to take place in the public space: artists are trained in and for the art world, and Romain Tardy finally anticipates that it will be taken by the big companies if the artists do not go to these other spaces.

4.2. A short history of projection mapping according to Dominique Moulon

4.2.1. *Projection mapping in the history of light*

Rather than considering projection mapping as a trend in itself, it seems appropriate to consider creations that are installed over time and that use light in space, whether they are interior prototypes or building facades. Without going back to Plato's cave, it is interesting to go back to the 18th Century with pre-cinematographic experiments, namely the magic lantern and the shadow theatre, the ghost paper, which consists of a complex device to do what today we would call holography. In these different historical cases, it is not strictly speaking a question of projection mapping, but of surprising and mysterious works inducing the use of light to give performances and make installations that question the public. Since the beginning of the Renaissance, the invention of the darkroom has allowed painters to compose their paintings using light installations where the image is inverted. The *camera obscura* is not of the projection mapping type, but it is nevertheless of the installation type where light has an essential role. With the invention of photography, then cinema and the video projector, history is experiencing more and more surprises over time and often larger and larger sizes. From the very beginning of experimental cinema, for example, a cinema of abstraction, a cinema of another form of narrative, a cinema of effects and surprises emerged. Similarly, creators will very early introduce projection on stage, such as the rock band Pink Floyd. Mike Leonard's Liquid Light Show was a kind of object cinema on stage in real-time, projecting surprising shapes. At about the same time, the beginning of video art produced works that questioned understanding.

4.2.2. The invention of the video projector

Video Jockeys, artists, performers used optical machines very early on, then the slide carousel. Since history of art is also made up of inventions and innovations, it then went on to experience a major one: the video projector. With him, new artistic practices emerge. Tony Oursler deploys it in institutions or galleries. Described as a video art, his work is in fact similar to the practice of projection mapping because he uses light to place images on puppets, spheres and other media. Similarly, Krzysztof Wodiczko uses video projectors, rather outdoors, for artistic and political practice. Projecting into the city when it is the place of politics allows the artist to integrate political issues into the urban environment with light. These two emblematic examples of projection on a medium indoors and outdoors, in small and large formats, show how video artists very early on took possession of the projector. In the 1990s, contemporary video artists built a discourse and discussion through this medium. The video projector literally offered the city as an artist's studio, just as the impressionists had already been pushed out of their workshops by the invention of photography and tube color.

Later, a strong craze for the video projector was felt by another generation of artists who invested inside and outside to create enhanced sculptures. For a *Nuit blanche* (Paris), the Spanish Pablo Valbuena proposed enlarging the façades of the Austerlitz station by redrawing them. There are other artists or collectives such as Anti-VJ in France, 1024 architecture or even Nonotak who practice light, even if it is not strictly projection mapping. In the 2000s, collectives set up internationally such as Graffiti Research Lab, which made screenings in New York, but which set the whole world on fire. They use lasers, projectors, open source tools and projects inspired by the open source culture.

Since then, practices with the video projector have become more democratic. Small models are nowadays very affordable, relative to the reach of all. It is also possible to rent or borrow them easily. For example, computers in the 1960s were unaffordable except for a few artists. In the 1970s and 1980s, the personal computer made tools available to everyone. It is for this reason that we see the concomitant emergence of video projectors and their diversion, the re-actualization of historical practices such as short-circuiting, but which are sometimes mixed with projection, possibly even on a support or on the ground. Paul Chan proposes a very interesting work in which he projects on the ground, becoming a support. Ryoji Ikeda makes large images on the wall and especially on the floor where the public literally lives. He is tempted to take off his shoes and invent a choreography inspired by the images.

4.2.3. *The feeling of immersion with different applications of projection mapping*

First of all, two interesting exhibitions (even if not everything is equal and not everything is art) take place at the *Grande Halle de la Villette* in Paris and are presented almost in parallel with projection mapping. Ryoji Ikeda's work deploys large images projected with radicality and discourse. The Teamlab collective brings a gentle*r* and narrative universe. Immediate rooms satisfy children while others can make adults think. The idea of creating an immersion is in the spirit of the times, to create images indoors as well as outdoors. However, the cost of these installations remains relatively high, unlike the projection of light into urban space, as seen at the *Fête des Lumières* in Lyon. In the continuity of the *Carrières de Lumières* in Les Baux-de-Provence, *Culturespaces* is offering a new space in Paris: the *Atelier des Lumières*. The projections presented here are iconoclastic because Gustave Klimt's work, for example, is literally cut and then glued back together, although we can suppose that collage already existed in the 1920s and 1930s. Nevertheless, appropriation seems more interesting when we talk about content that is not necessarily artistic or amateur. In the same way, it is interesting to look at the practice, the post-Internet trend by the generation of 30/40 year olds who are now in galleries. We can see that electronic, technological and digital art has moved from the festival to today's institutions, galleries and major events. In parallel, there are, as with the *Atelier des Lumières*, events that have more to do with the continuity of yesterday's sound and light shows on castles, churches, etc. It is a tradition of cultural events that continues rather than new artistic proposals. The fact is that the projector associated with computers allows graphic designers as well as companies to use brand names in public spaces or in interiors. This allows both emerging and established contemporary artists, companies and cultural structures to generate installations. A certain mistrust of projection mapping practices results from this on the part of purists of contemporary art since projection mapping is practised extensively. The software is extremely affordable, there are even free ones, collectives abound, local authorities and brands want more or less projection mapping works for their town hall or festival of small or large size, with light or video projector.

By referring to Olafur Eliasson's *The Weather Project* (2003), a major work of the 20th Century, we realize that this reconstituted sun is not projection mapping either, but a work of light. However, we find the same idea of excess in the use of light by an artist in a studio who is used to collaborating with engineers, physicists, theorists, graphic designers, architects, urban planners, etc. In the same way, when we talk about these contemporary practices inducing the jet of controlled light with computers, we find ourselves with a whole range of creations that go from interesting cultural phenomena to be analyzed, that can be in good taste, or events that are at the service of brands and for an audience trying to surprise them. Some of the works are quite interesting, which leads us to continue to see them. What is interesting is the fact of mixing the use of the video projector with the possible interaction, the possible generativity inducing public participation. History therefore brings together major artists of an adventurous contemporary art offering exceptional pieces.

In parallel, we find technology in scenography with dance, theatre, or visual performance. In 2018, the Indochine band's concerts installed a ceiling of light above the stage. Companies like Moment Factory (Montreal) are used to take over the stages of singer Madonna and other great contemporary stars. The fact that these tools, techniques, practices, trends and uses are used with talent by companies as well as in art, for the cultural world, including scenography, or even politics when we intervene in the urban environment to seduce voters, so they can escape the categories.

Today, companies of graphic designers, creators or artists do scenography with light, with video projectors by making images and texts coincide with very particular areas of the environment. In addition, museums sometimes take an interest in digital as a subject. For example, the exhibitions "Electronic superhighway" by Nam June Paik (Whitechapel, London), "Art In The Age Of The Internet" (Institute of Contemporary Art, Boston) and "I was raised on the Internet" (Museum of Contemporary Art, Chicago) introduce digital as a subject in contemporary art and in major institutions. Small, large and medium-sized institutions integrate digital technology into their museology: projection mapping, but also touch tablets, projectors, augmented reality, virtual reality, the drone, the led for its new possibilities offered by addressable controlled light.

4.2.4. *The role of ICTs today and tomorrow*

Revealing the specificities of a tool, a medium, a digital subject allows us in the art to question and give keys to reading the society in which we live. The video projector, for all that it allows, enables us to enter into a large number of artistic practices ranging from intimate pieces in galleries to monumental pieces in urban spaces and many other fields invested such as live performance. In this ballet of curiosities, the tools can sometimes be found at the service of brands or politicians – with nevertheless sometimes interesting proposals from activists. This makes it possible in a few minutes to increase the city with words that can be resolutely committed. The video projector has given rise to many practices. In Madrid, in 2015, for example, hologram demonstrations were organized. Opponents of the country's new internal security law marched in the form of a hologram. All this is so rich that the public can get lost in it. The video projector or the digital projector as a whole is at the same time a tool, a medium and a culture that is now accessible to everyone and that holds surprises for us.

The 1990s were a time of virtual delirium. The slightest immersion questioned the public, who were attached to the scanner object, which made it possible to virtualize the world. Since then, we have been living in a resolutely digital world, perhaps to the point of wanting to rematerialize it, if only in fragments. That's why the 3D printer flourished with the idea of putting in the palm of his hand what you could previously only have in front of his eyes, on a screen. This idea of materialization is also very visible in the gallery when the lack of access to non-permanent, virtual, digital, interactive, complex works, with computers, etc., has been a problem in the market. However, digital has developed in contemporary art and at the same time in festivals such as Ars Electronica (Austria), I Love Transmedia (Paris), Némo (Paris), Seconde Nature (Aix-en-Provence) and elsewhere. Today, there is the generation of post-Internet artists who refuse to be named trends in the same way as the artists mentioned above who do not consider themselves projection mappers even if they are not very far from it. Although there are works on the Internet, in the *White tube* of contemporary art to date, images from the Internet are printed, virtual content is extracted and rematerialized by hanging them on walls, putting them on the floor, hanging them to make sculptures, printing them in 3D or via laser or waterjet cutting. Today, great contemporary artists such as Barry X Ball or Jaume Plensa use digitally controlled machines to do what they consider sculpture in the digital age. Prototyping via digital machining allows them to achieve a form of perfection, which is much more difficult to achieve with a chisel.

We can see that it is apparently through sound and image that artists have discovered the power of electricity, electronics, computers and digital technology in general. Perhaps the world of sculpture has naturally been spared the extent of digital technology. This is not the case today. The book *L'art au-delà du digital* (2018) defends the idea that from now on, although one is an oil painter, one will not start a work without an internet search. However, we know that search engine algorithms format our thoughts to some extent. The painter, sculptor, architect... all today are imbued with the soft power of San Francisco, the power to inspire us as well as to format us. Algorithms are part of the guidance of our thoughts, our memories and even our "artificial" intelligences. What illustrator, painter or sculptor could today claim never to be influenced by the world in which he lives? This is why it seems interesting to consider digital as a tool, as a medium or as a subject. Sometimes all three of them at once, but never any of them. To quote Norbert Hillaire (2009), we should now consider the "digital coefficient" of works rather than talking about art that is digital or that is not. All the works are more or less digital. Hillaire borrows from Marcel Duchamp his "coefficient of art" which envisaged that the objects also had a coefficient of art that the artist revealed.

4.3. Projection mapping and its limits according to Christiane Paul

4.3.1. *The New Aesthetic*

The aesthetics of digital media is a topic Christiane Paul has been involved in since the late 1980s, tracing its different movements and developments. The idea of the New Aesthetic, which was first outlined by James Bridle in 2011, captures the embeddedness of the digital in the objects, images, and structures we encounter on a daily basis and the way we understand ourselves in relation to them. The New Aesthetic is a blurry concept or, to use a term by Hito Steyerl, a ghost of an image — a visual idea in the making but at the same time very valuable because it makes a statement about its own condition. The achievement of the New Aesthetic and perhaps the reason why it became such a meme is that it captures an important quality of aesthetics right now, yet it does not succeed in creating a framework for an in-depth understanding of aesthetics. This was one of the motivations for Christiane Paul and Malcolm Levy to look at the genealogies of that New Aesthetic, many aspects of which have a fifty-year history. The current discussions surrounding post-digital, post-Internet, post-medium work are closely related to the New Aesthetic because they struggle with similar issues, particularly the relationship between networked technologies and the object. All of these terms ultimately describe a condition of artworks deeply influenced by digital technologies on various levels but not necessarily taking digital form. They manifest as paintings or as sculptures but could not be understood without a deeper understanding of digital technologies and their aesthetics.

4.3.2. *Projection mapping as a technology*

Projection mapping is a projection technology that uses objects or spaces, as irregularly shaped as they may be, as projection surfaces. Depending on how you define projection mapping, its practice goes back a long time and has many predecessors. People may have tried to achieve effects similar to projection mapping centuries ago taking advantage of whatever was available to them at the time as an object, light source, and projection tool. Projection mapping as a projection technology in the narrower sense has also been used for decades. Michael Naimark, for example, employed it in an installation titled "Displacements" (1980–84), shown at the San Francisco Museum of Modern Art in 1984. But it has also been used in the industry or entertainment, an early example being the Haunted *Mansion* ride in Disneyland (1969). Projection mapping has had a long history but, as if often happens with technological forms, it takes a while until the technology has reached a level where it can unfold its potential. The digital age has the potential to bridge various gaps between art, science, and technology, which have always intersected to varying degrees. The Leonardo journal, published by the MIT Press, has chronicled the application of contemporary science and technology to the arts and music since 1968. Digital technology lends itself to a very broad spectrum of practices, which is something that distinguishes it from previous technologies. Projection mapping, as part of these technological practices, may not be new but has reached a new level of application today.

4.3.3. *Projection mapping as an experience connecting the physical and the virtual*

Projection mapping, virtual reality, and augmented reality are not movements in the traditional sense, but different ways of applying technologies. There is a certain overlap between the concept and aesthetics of AR and projection mapping because both of them try to enhance the physical environment through a projection or virtual overlay. Virtual reality is a very different category since the spectator becomes completely immersed into a virtual world separated from the physical one. In the case of projection mapping and AR, spectators experience close relationships between the environment surrounding them, the objects within it, and the immaterial overlaid digital images. When it comes to the experience of technology, we need to analyze the intersection between the virtual and the physical and how our senses perceive it. There are huge differences between the ways in which our senses engage with browsing the "virtual space" of the Internet on a computer; or being immersed in a virtual reality world; or having virtual elements overlaid onto the physical world; or engaging with the virtual through phones, tablets, and networked devices, from watches to bracelets. All of these degrees of immersion, as well as the respective interfaces of the technological devices supporting them are perceived differently by our senses and have distinct visual, aural, and haptic effects. Levels of interaction and agency are yet another important aspect in the experience of technologies. We have used technologies to extend and visualize our bodies for centuries, from microscopes and telescopes to technologies of representation such as photography and film. The body's relationship to technologies is always very active and there are two main aspects to it: the "enacted" body as our embodied experience of the world, the way we position ourselves and our mobility within a given environment; and then the represented body, the images we create of our bodies. Both are affected by technological evolution. Immersive technologies – from the early three-screen Polyvision format used by Abel Gance in his 1927 film Napoleon to the Oculus Rift – change our embodied experience, and imaging technologies influence the way we represent ourselves. Digital media translate the notion of 3D space into the virtual realm and thus open up new dimensions for relations between form and space: they renew the notion of sculpting.

4.3.4. *Projection mapping and museums or art institutions*

Art or its concepts should never be forced to conform to a space. Digital works, in particular, often have to be adapted to a particular presentation space, but there might be video mapping works or installations that simply can't be shown in a room or in a gallery. In "Displacements" Naimark transformed one room that can be replicated or installed in a museum environment. Other projection mapping pieces, such as several ones by Krzysztof Wodiczko, are meant to transform a specific statue or a building and can only live in that context. Galleries or museums are able to present a piece like that in public space and realize a public site-specific installation to create new spaces for art. Projection mapping does not have to conform to the modernist white cube, originally configured to accommodate the presentation of static artworks. Digital practices often require that museums and galleries expand their customary methods of presentation and documentation, as well as their approach to collection and preservation. Today institutions are becoming much more flexible in accommodating time-based, dynamic, participatory, customizable, and variable art forms in their presentation formats. Projection mapping is a technological form that is more commonly used within digital art practice.

Christiane Paul hasn't worked with projection mapping as part of a project shown at the Whitney Museum yet, but might do so in the future in ways driven by the conceptual use of the technology, rather than the showcasing of technology itself.

PART 2

Texts and Techniques

5

Listening to Creators in Residence

5.1. Creators, a residence and a festival

Programs, computers and projectors are the daily environment for contemporary projection mapping creators. Just like painters, sculptors and artists in general, the skills they deploy require long-term learning, at least as much as they depend on a particular disposition to be created. The medieval apprentice painter lives with his master for five to six years and the first tasks he is initiated in are primarily material. As Pascale Bédard (2014, p. 51) says: "To be an artist is to practise a profession, that is, to acquire and put into practice skills in the accomplishment of a particular manufacture, that of a material or performative art object." Thus, 17th Century Dutch painting frequently depicts the artist's daily work in his studio as hard work. The painter is represented at work, with his tools: palette, brushes, knife, hand support, etc. However, David Hockney (2001) points out that art history frequently mobilizes faith in the artistic gift, which reduces our ability to think about the artist's work in its multiple dimensions. In our case, the training in the practice and tools specific to projection mapping is long and complex. It requires patience from the creator to reach the knowledge of a tool and the ease of its practice.

As part of the first edition of the Video Mapping Festival, thirteen creators in residence at the Arenberg Creative Mine site produced works that were broadcast during an evening in March 2018 in Lille. These residencies were an opportunity to conduct research on the creative experience, guided by the following questions: can we access the creative dynamic experienced

---

Chapter written by Marine THÉBAULT and Daniel SCHMITT.

by creators? If so, what can we observe and understand about this creative dynamic? What perspectives does this understanding offer? By studying the intimate experiences of creators during their production activity, we question the modalities of creation in residence.

5.2. Capturing the genesis of a work

How can we capture and determine the genesis of a work? The genetic criticism developed in French literature in the 1970s attempts to "track down the future of an artwork by studying the written traces of its genesis" (Grésillon 2016, pp. 7–15). The researcher makes "hypotheses on scriptural operations" in an attempt to "guess, detect, deconstruct and reconstruct the *paths of creation*" from the manuscript archives accessible to him. This is also what historians are trying to do by collecting corpora as exhaustive as possible to interpret them. Sometimes, contemporary writers such as Jean-Philippe Toussaint and François Bon post numerous paratexts on the Internet so that we may have the impression that we know precisely the genesis of their works. Nevertheless, these deposits form new creations that are added to the first without revealing everything. So, is it useful to collect an embodied memory of the artwork, a memory told by its author to capture the genesis of an artwork, the meaning that the author gives to his actions and choices during the creation? And does this memory reflect the environment, the activity and emotions that accompany the creative process?

The creative experience described in the first person by its author could help to understand the sensitive ecosystem in which the artist's body, trace and related environment interacted. However, asking someone to describe aloud their immediate experience engages them in a reflective analysis of their activity. Varela, Thompson and Rosch (1993, p. 32) reacall: "By meditating or becoming mindful and aware, one is disrupting one's normal mode of being in the world, one's active involvement and one's taken for granted sense of the world's independent reality". Other methods such as the elicitation interview (Vermersch 1994) make it possible to question the artist *a posteriori* and to obtain highly refined verbalizations. Nevertheless, the longer the experience, the more difficult it becomes to remember every moment of your experience. The REMIND method (Revivification, Experience, Emotions, Sense Making micro Dynamics), proposed by Daniel Schmitt and Olivier Aubert (2016), partly addresses these issues. Its implementation with creators in residence allows access to the dynamics of sense-making during the act of creation.

5.3. REMIND: a method to capture the dynamics of the situated creative experience

The REMIND method (Schmitt and Aubert 2016) is mainly based on the theory of enaction (Varela *et al.* 1993) and the theory of the course of action developed by Jacques Theureau (2006). This survey method uses the recording of the subjective video trace of the activity of the creators surveyed to stimulate their subsequent recollection. We equip a creator with an eye-tracker to record his subjective visual and auditory activity. This method makes it possible to reconstruct the body, cognitive and emotional situated dynamics of the creators' experience. We have programmed these surveys in the form of coring in the time of creation of the residents. They were equipped for an average of one hour at three different stages of their creation. The creator lives his experience in an almost natural situation, without the presence of the researcher. Following the equipped activity, the creator is placed in front of his subjective video and we invite him to describe, comment and tell his activity. We stimulate his capacity for recollection to access his pre-reflective consciousness (Theureau 2006) by avoiding a *posteriori* analysis or an explanation of what would not have been experienced. What guides us is to approach the underlying meaning of the action. We have enriched Theureau's analytical framework with hedonic valence to identify the creator's state of pleasure and displeasure in the course of his action. This framework makes it possible to identify the elementary sequences of the meaningful activity from the creator's point of view. Through the creator's stimulated verbalizations, we identify 1) what he takes into account in his representation of the world, 2) what he does, his commitment, 3) what his beliefs are and the knowledge he mobilizes, 4) what expectations are at stake and finally 5) the qualification and situation of his emotional value on a scale ranging from -3 (displeasure) to +3 (pleasure).

During the winter of 2018, we conducted 21 REMIND-type surveys of thirteen creators from eight different countries. Valeria Amoretti, anthropologist and archaeologist, and Donato Maniello, projection mapping artist (Italy), enhanced the Arenberg mining site through a video projection on a model of the site. Dimitry Bulnygin (Russia) questioned our everyday routines on the walls of the salle des pas-perdus at Lille-Flandres station.

Olivier Cavalié and Gervaise Duchaussoy (France) animated the rue de Béthune with a series of micromappings playing with the architecture of the shopping street. Maeva Jacques and Claudia Cortes Espejo (Belgium) proposed a projection mapping in the underground of the Saint-Pierre Canal: monsters and mysterious characters tell the story of this medieval vestige. Joan Nieto and Javier Canal (Spain) decorated the Republic Square with fractals inspired by the sounds they produced. Robert Seidel (Germany) has covered the architecture of the atrium of the Palais des Beaux-Arts with moving abstract images from digital analog image creation techniques. Thomas Voillaume (France) projected a projection mapping on a four-meter high sculpted character at the Ilôt Comtesse. Paulina Zybinska (Poland) and Jelle Van Meerendonk (Netherlands) told the story of the headquarters of the daily newspaper *La Voix du Nord* like an animated film. With their agreement, we have kept the first names of the creators.

Figure 5.1. *The resident is equipped with an eye-tracker during a work session (source: Chloé Rougier). For a color version of the figures in this book see, www.iste.co.uk/schmitt/image.zip*

5.4. Space, tool and solitude

Based on the transcription of the 32 hours of verbatims collected, we identified the different key sequences shared by all the creators using the Advene software (Aubert and Prié 2005). During the first days of the residency, the creators settled in what would become their working environment. The workshop is understood as the strategic place to start the creative process. As a physical, communicative and potentially action-oriented space, it is favorable for creation on the condition that its creator feels comfortable there. The organization of the creator's practice through the implementation of technical tools shows that residents have encountered various technical obstacles throughout their creative process. Do they constitute obstacles or, on the contrary, crossroads in the creative process?

Do creators understand the pitfalls in a similar way? How do creators identify and circumvent them? Are they developing strategies? Can these problem-solving methods be categorized?

5.4.1. *The instrumental space*

The workshop is a production center and the following examples illustrate what it represents for creators. Robert points out that the time it takes to adapt to the workplace affects his work method and his cognitive state. For residents, this time can be both inspiring and rewarding as well as a time of frustration and discomfort. Robert uses the sunlight in the room where he works as a new tool to cast shadows and "instantly create a new drawing". He also finds it "interesting to see things from two weeks ago as part of the reflection and references". Nevertheless, the table in his workspace is too small and having to store his pencils interrupts his activity. He would like to have a large artist's studio where there would be a table for the computer, one for pencils, one for colors, another for cutting paper. For Robert, as for most other creators, the residence as it was organized in Arenberg during its first edition was not sufficiently inspiring: it was too isolated from other cultural and commercial spaces. Especially in the evening, the creators felt trapped in a kind of no man's land.

5.4.1.1. *Preparation, warm-up, planning*

The working environment includes all the material conditions, services and resources made available to the resident for the performance of the task he wishes to accomplish. Projection mapping is "a technique that consists of digitally mapping a physical place in order to be able to apply projections appropriate to its relief" (Rautureau 2014). The preparatory work of writing the intentions and learning the projection context seemed important for creators to organize the technical needs related to their activity. Gervaise pencils, sketches and notes the intentions to establish a well-organized planning, precise specifications. Similarly, Paulina prefers to have a plan prepared in advance, to know what she will do from start to finish with Jelle, her teammate. During the first week of residency, Jelle wants to define the concept quickly, have a very clear content in order to reserve as much time as possible for the creation. According to him, it is not only a question of creating, but also of thinking logically. He and Paulina undertook to "discuss the right techniques to properly calculate the concept 100 %". Robert, who has a plastic practice, prepares the gesture with his hand by repeating several times the curves of the drawing he wishes to obtain by twisting a sheet of paper. When he feels ready, Robert launches the video recording of his movements, both a trace of his creative process and the material to create the projection mapping. According to him:

> "You have to be ready otherwise you capture hours of material that will never be used. Maybe there will only be one minute that will be important and yet we will have to look at everything. For this type of process, it is necessary to have a specific control of the hand to try to make yourself as invisible as possible while preserving the quality of the line, the quality of the tool."

To prepare their technical actions, most creators need to know very early and accurately the location and object of the projection. They sometimes lack this information. Claudia and Maeva, for example, do "things where there will be no need for the size of the set" because they lack the template. However, they cannot think and progress in the animatic without it. In the same way, Paulina tries to know if the building has already been mapped, if there are masks, a 3D model or a model: "It is very important to have them as soon as possible because otherwise it is difficult to start the animation." They need precise information on architecture. Knowing these details about the shape of the projection surface, the dimension, the texture, the model, the place of the spectator, etc., allows creators to make creative choices as early as possible. For example, Olivier had an idea for a projection that he liked, but once he was in the field, he realized that it would not be real. It was on location that he was able to measure the constraints of the site. Similarly, Jelle and Paulina want to make the spectators appear in their projection mapping. It turned out to be complex because they would have needed a lot of light, but they realized during the spot that it would be very dark at the time of the projection. Robert likes to draw fine lines, but considering the spectator's distance, he decides to draw thick lines so that they are clearly visible during the projection.

While being aware of the projection area and its requirements, the creators install the material resources they need to apply their work method. The challenge is both to choose and master tools adapted to their needs. According to Pierre Rabardel (1995, p. 23), "a technique only exists when it is practiced, that is, when it passes through someone who, having learned or invented it, implements it effectively". Skill reassures Olivier: "I will master it, so it doesn't scare me." Residents selected tools and conducted a series of tests to understand, remember or try out applications. According to Gervaise, the tool is "chosen in relation to the intention of the project". During the residency, she chose to work with the TVPaint animation software. She starts by doing a first import test while looking for the commands of the animation software because she uses it less than others, which makes her practice less fluid. At first, Gervaise "feels her way around, organizes herself, puts ideas back in place". When Claudia directly *cleans* the drawing, she realizes that she is doing "really crazy things", but she needs this step to warm up, to "get her head in order". Similarly, Paulina does not have the software she is used to, she adapts to another software she does not know, but she takes time to understand its interface and parameters. For his part, Joan looks at some tutorials as he usually does to see how he could use the new version of an After Effects plugin that he never used for his project with Javier. To understand the plugin using the tutorial, Joan tests parameters like velocity and light to see how the object changes. Robert, who has not drawn for a while, tries his pencils, which engages him in a process of relearning, of recalling the gesture. Finally, using the Toon Boom animation program, Jelle draws shapes commonly used in animation and reproduces an animation he has in mind for having often made it. According to him, it is then a technical and routine task since he has already done it many times, he has the experience. These examples show how creators select the tools they consider necessary to carry out their project. Some use the tools and processes they usually master, they choose to use software that can be complex, but that they know perfectly well, to which they are accustomed (Pignier 2012). Others (re)learn a practice during their act of creation.

5.4.1.2. *Irritating constraints*

5.4.1.2.1. Computers and programs

During this phase of technical development, obstacles may require additional cognitive effort on the part of the creators. In some cases, they have to use different computers with different technical characteristics and different software. However, having to change workstations can interrupt and compromise the creation process. For example, Gervaise wants to work on a template, but she realizes that she has not downloaded it and that there is no internet access on the computer on which she started working. She has to move from one to the other, which makes her activity choppy. Similarly, Jelle changes computers to animate part of her video because the compositing software is installed on the laptop while the animation software is on the desktop computer. On the other hand, Paulina uses software that she has never used before because the license of the one she usually uses has expired. Paulina wants to renew it later, especially since she owns a Mac while in her residence she is on Windows, so even if she bought the license, she couldn't transfer the software to Mac.

Creators have reported strong computer constraints that they believe hinder the creative process. At the beginning of a recording, Dimitry waits for the project to open in Final Cut Pro, while he checks his messages on his bank's website. Later, since it takes time to render a sequence of the timeline of his project, Dimitry starts looking out the window. There is "nothing else he can do". He then watches something on Facebook, the weather forecast for the day of the festival and the results of the presidential election in Russia. Paulina is in the same situation. According to her, "rendering takes time, it's long to preview because these computers are not the fastest, so you have to compose in twice as much time, it's a lot of time". Robert wants to import photos into the software he develops himself, but before that he has to turn the computer back on because of a "stupid network story" he doesn't understand. Then, it must rotate vertical videos 90 degrees. According to him:

> "By doing everything as simple as possible to make people know how to use a computer, it becomes more complicated for them because they get things they didn't want and have to change them. It's totally stupid because some software doesn't see where you're going and others do it automatically, it's extremely confusing."

Then Robert has to turn the computer back on because it is overloaded, but it takes some time before it is operational. When the computer restarts, the screen remains blank for a long time. The machine sets the rhythm of the creative time, it imposes its rhythm and this frequently conflicts with the desired activity of the creators.

5.4.1.2.2. Projection mapping expertise

Among the creators, those who have not practised projection mapping have encountered many difficulties. For Gervaise, who comes from animation, the choice of the format for importing an image worries her: "This is projection mapping, it's different for me because the format of the films is already established and I work in it whereas this is another format,

I'm wondering how to proceed." Later, she explains that part of the building she will map is glazed, but until it is projected, she doesn't really know what it will look like, "there's a little suspense". Besides, Paulina "has never really worked in animation, it's always 2D animation, so it's different". According to her, things can look good on the screen, but once projected, there can be a framing difference between the computer screen and the building, for example. Paulina is "a little afraid to see what will really work or not on the building". Claudia points out that animation is a world apart, it is not easy to know how to break down movement. In this way, as she tries to animate characters, she realizes that her state of mind is only in illustration, which she has been practicing for several days, which prevents her from proceeding correctly. She "totally lost the framework of animation".

5.4.1.2.3. Creativity in creation

Ignoring details to save time

Claudia could spend a lot of time working on the details of an illustration, but she prefers to have a quick result that works and not "get bored by anything". Similarly, Maeva uses the TVPaint software and proceeds directly to *clean up* by thinking that the deformations will not be annoying. She designs guitars without strictly respecting their shape because they have only four strings and makes "a somewhat disgusting connection of the strings". She thinks it's not embarrassing because on the brick you won't see the fine details that inform you that it's not a guitar: "In animation, if you see that it passes, you don't take too long to do the things exactly." On the interactive part of the project, as Paulina readjusts some of the renderings, she realizes that one of them is not perfectly animated: the main character's jacket "passes a little bit through the skin". Nevertheless, Paulina believes that this is sufficient for this part because it is a very difficult process and it takes a lot of time to do it properly. She then notices that the movement is very slow and says to herself that "the result is not the best it can be, but it is already better than before" and leaves it as it is.

Bypassing the problem by changing tools or materials

Paulina and Jelle are trying to translate a French article into Dutch, but the Google Translation application they use is not working well. The application takes time to calibrate the text properly and offers a partial translation that makes it difficult to grasp the meaning of the text. Finally,

Paulina decided to do some research on Wikipedia. For his part, Robert plays with his drawings by exposing them to sunlight to obtain a new rendering through movement and shadow. He notices that the weather is unstable, which leads him to look for a lamp with the same color temperature as the sun to be able to continue his work. Thomas discovers by chance that the polyurethane foam with which he filled his wooden sculpture is flammable: it burns while he welds next to his sculpture. He is upset because he wants to make a dismountable and light sculpture to present it in festivals. He thought he should have filled the sculpture with another foam classified as fire, but he didn't have time to get it. He is looking for ways to fireproof his room, but it is complicated because the wood must be fireproofed on both sides. He researched other materials such as resin and learned that it was also flammable. He still decides to use it and goes to the DIY store to get it. Finally, from an archive that deals with the creation of the Arenberg mine, Donato tries to extract text to apply a video effect to it. The extraction operation done using Photoshop software is difficult, so he chooses to extract the text and redraw each letter using Illustrator instead of deleting the background of the document in too low a resolution with Photoshop.

Renouncing an action or idea

Gervaise wants to recreate the template from which she works using Photoshop software. She then decided that it would be preferable to use the last photograph sent by the Director of Rencontres Audiovisuelles, an association promoting projection mapping. She retrieves it and retouches it and realizes that the ratio of the video has changed. Gervaise reformats the file to be able to import it at the right ratio. Finally, she decided to use her own photography to create the template. She believed maybe she should have thought about her strategy before she started and had to think by doing. While Robert tries graphite on transparent paper, it doesn't work well, he gives up the idea. Similarly, he uses a blue pencil to add structure to his drawing, but none of the colored pencils Robert owns in residence are suitable for transparent paper. Upset, Robert has to create a black and white projection mapping. Joan is looking for a tutorial about adjusting the light intensity. After a long search on the Internet, he finally found one that allowed him to adjust the intensity as he wished. However, he finds the manipulation too complicated and decides to put this idea aside. Similarly, Paulina and Jelle put aside some elements of their storyboard when they realized that the final animation would take too long. When Dimitry generates the rendering of his work, he realizes that it is the wrong background that appears. He doesn't understand why the last pictures are at the beginning when normally the camera automatically numbers each picture and puts them in the timeline ordered by date. It "drives him crazy" and he decides to give up for now and go back to what he started doing.

The importance of the workspace

Through the residents' testimonies, there is a strong need for a cognitive and physical environment conducive to creation. All the actors of the festival share this idea, but what seems obvious to organizers, researchers, software developers, etc., is not necessarily so for creators. Not only can a third party imagine that the environment is adapted to the creator, but also underestimate the time it takes to adapt to his workspace and then to accomplish his task. If we listen to residents, the inadequacy of the environment and resources slows down the creative process and is accompanied by a negative feeling. Creators want to have a personalized creative space that meets their needs and uses. Bédard's research (2014, p. 167) confirms this: "A quality project requires adequate means of production, i.e. an appropriate location and tools." If the creator considers that the place of residence is outlying, that the layout of his temporary workshop is not adapted to his needs or that he does not have the necessary means for his production, then his activity and his well-being will be altered.

Beyond the need for a personalized space, the creators expressed the wish to have very early on precise information on the object and the place of projection at the risk of slowing down or even stopping the creative process. Due to a lack of details on their projection object, residents were forced to wait to move forward and redirect the planned work accordingly. The creators would like to have the templates, models, masks, etc., well in advance of the residence to guide them in their technical and creative choices. This is all the more so as the choice of tools and their handling has proved to be an important, delicate, laborious and hesitant subject of concern. Residents managed most of what they needed independently, sometimes at the risk of feeling isolated. The interviews report on how creators have resisted, negotiated, diverted or bypassed problem situations that are often linked to a tool. By wanting to solve what they perceived as dysfunctions, creators had the feeling of dispersing themselves to the detriment of their creative work. This feeling reinforces Bédard's (2014) approach that creative activity is more about work and training than inspiration and expression. To compensate for their lack of knowledge of the tools, creators are looking for alternative solutions. Tactics are sometimes based on the creator's experience (not paying attention to details saves time), sometimes on his teammate or an external resource.

When the user fails to grab a tool to make an idea a reality, then that tool becomes limiting. Remarkably, the idea is often guided by the mastery of the tool: sometimes when an application is extremely constraining and makes the design task problematic, creators have finally abandoned the very use of the application or the realization of their idea. Creators have each faced these pitfalls in their own way by deploying different strategies, but in most cases they have led to the abandonment of an idea or activity. Let us be clear: this is not the so-called "essential uncertainty" for conducting innovative artistic work (Menger 2009). On the contrary, this lack of knowledge of the tools seems to have worried creators a lot without having any added value on the effectiveness of their work. From their point of view, this time spent alone arbitrating material problems did not have an enriching influence on the creative process. This is why we distinguish between the structuring constraints that favor creativity in certain cases and the irritating constraints that monopolize the creator and keep him away from his object of creation. Various material obstacles have led the creator to ask himself questions and have doubts during his creative experience. He is slowed down because he consumes part of his residence time learning to handle the tools on his own. This double isolation linked to the geographical and cognitive distance from the place of residence, but also to the lack of technical support, seemed cumbersome and counterproductive for the residents. From the creators' point of view, when they have not succeeded in resingularizing the artefact and making it become an instrument (Rabardel 1995, p. 6), these difficulties have constituted limits more than they have helped the creation or have been support for imagination.

On the other hand, residents seemed to need accurate knowledge of projection mapping tools. The Fine Arts Schools were created to transmit a know-how offering everyone the means to excel in their art. In the same vein, a team of supervisors seems necessary to meet the needs of creators who may feel alone in many problem situations. Still, from the creators' point of view, being accompanied by a competent technician would significantly improve the conditions for creation. It also seems that support, as an artistic director does, in the permanent construction of a personal work space and in the creator's gesture is necessary, especially for novice creators.

5.4.2. *The dynamics of the emotional states of the creators* in situ

The REMIND method gives access to the emotional states of the creators. During the stimulated recall interview, the creator regularly indicates his feelings on a graduated scale from -3 (dissatisfaction) to +3 (satisfaction). We observed many negative-valent sequences during the creation process, regardless of the reception conditions. Robert means that part of the reality of the author's work is not interesting. In addition, he saw his projection project at the Palais des Beaux-Arts in Lille with a lot of pressure and tension in his body: to create something interesting for "such a great museum". For Maeva, drawing and putting things in order does not require reflection, it is automatic and it is not "fun" (0). Although she prefers to do animations and transitions, her emotional state does not change when she gets started (0). Similarly, her teammate Claudia finds the facilitation process "boring". Paulina sometimes feels her creative process – for example, animating a projection mapping scene and seeing how it looks – as something "so boring" because it's very easy to do. She can then easily share her mind between two or three different activities that she carries out simultaneously. For his part, Thomas asserts that artistic work is daunting and slow. Mounting his wooden sculpture is repetitive, boring (-0.5), does not require reflection, can dissatisfy him (-2) and requires courage. He hopes several times that someone will come and talk to him to entertain him. Finally, Dimitry believes that work can never be perfect and he is never completely happy when he works.

The preparatory phases of the creative work are very often charged with a negative hedonic valence. For example, the sorting phase of his experiments and making decisions annoys, irritates and stresses Robert (-1, 0). He finds it neither creative nor interesting. In addition, it requires him to make compromises between the many ideas he has and what will work for the projection. He would be happy to be able to only experiment, never to finish anything. But Robert points out that we rarely spend enough time doing these kinds of things because there are so many other things to do. Donato, on the other hand, does not like to extract elements from a document (-2.5). He hates and is bored organizing files (-2) because he considers that it is routine and, in this sense, that it is neither artistic nor creative and that it is an imaginary phase of work that is not concrete. Gervaise does not like to manage what is part of the technical organization, she finds the implementation agitated. Before starting to create images, she must import her template into Photoshop, then adjust the truncated image and set it in TVPaint, it's long (-2). Javier doesn't find it exciting to design a 3D model of the buildings he will map.

Second, residents report negative emotions when the tools resist them. Robert regrets having set his software for 45 minutes and five times to have only 10 minutes during the day to improvise with his drawings. Moreover, he is confronted with technical problems that he knows how to solve, but he finds that "it is disappointing when you have a vision, an idea, when you put everything in place, you have to accept it". Robert is really annoyed not to have transparent paper pencils with the same qualities as those that work on white paper. Claudia feels "really bad" (-1.5) when she realizes that she is not using the tool properly. When Dimitry renders and discovers that the structure is strange and the wrong background appears, he does not understand where this anomaly comes from, which makes him "crazy" as we saw earlier. In addition, he works with an older version of Final Cut Pro that seems sufficient for his work, but some effects in this version do not allow him to do what he wants. For example, Dimitry finds himself forced to render his work as if it were a film, which is complicated. Donato hates the multi-tasking aspect imposed by technology (-3), he sometimes plagues against the software he uses.

On the contrary, creators show a real pleasure (positive hedonic valence) when it comes to "pure creation", as Gervaise points out. Once she has installed and managed all the necessary elements, she begins to make the contours of the building, which gives her pleasure (0, +1). When she draws, it is more stimulating (+1, +2). When Robert does the automatic writing, he also experiences positive emotions. If he sees something interesting emerging, then he's excited. Similarly, when Jelle and Paulina have creative ideas that emerge from the source texts they read, Jelle feels good (+1) because they are "creating something very good". In the same sequence, Paulina is excited (+1) because they are "on track". Olivier takes pleasure in experimenting on a computer, in letting himself go: "Now I'm good, I'm having fun" (+1, +2). Thomas enjoys tinkering with a tangible object, using all his senses. Dealing with the Première software is a phase of the work that Donato loves. According to Dimitry, creativity is about gathering elements, animating them and then editing them. Mastering the tool seems essential: Claudia is happy to animate correctly, to have understood how to do it (+2, +2.5). She then feels very comfortable in what she is doing.

Achieving the desired creation, then seeing it rendered, is really a source of satisfaction. Dimitry stresses that he is only happy when the work is fully completed (+2). When Paulina realizes that 95% of the animation is done, she feels less stressed. When Javier sees the result of his work, he also goes up in the scale (+0.5). Donato is eager to test the final result on the model, which leads him to put aside his activity to immediately design an idea under After Effects and see the effect of the visual a few days later. It is only when Maeva sees the result, "that it moves", that she is in a positive state (+2). Even watching other people's animations moves Joan up the ladder: he loves watching the animations of the different tutorials on which he is working (+3).

5.4.3. *Work, emotions and troubles*

The residents have demonstrated a variety of hedonic valences during their creative process. They have had different, intense and varied experiences. The intensity of the most negative emotions (between 0 and -3) corresponds to the creative process in the broadest sense. Residents did not feel pleasure, contentment during most of their creative process. This is an obstacle to the idea of a pleasant and happy artistic activity, inspired by a higher authority, which would give rise to a work of almost magical status. As Bédard (2014) points out: "Artists themselves are much less romantic about their own situation than the general public seems to be." Preparing and adapting to technical tools does not make creators happy. Nicole Pignier (2012) also notes that most creators experience negative emotions when interacting with digital tools that resist them. We also observe that creators make a separation between the intelligible of computer programs and the sensitive of artistic forms. The artefacts that everyone tries to appropriate, transform into an instrument and organization throughout the creative process involve convolutions and struggles accompanied by negative feelings that occupy the creative space in an important way. They show a positive hedonic valence when it comes to creating images, experimenting, tinkering, composing, gathering elements, animating, mastering the tool, feeling comfortable and on track. Finally, test the final result to see the effect, even look at other people's animations, move them up positively on the scale. Through these struggles, moving images are born until the final result (positive valence). Episodically, they have experienced states of satisfaction, especially when they have achieved their goals. The tool, whatever it is, is important in the creative process and its control ensures a minimum of comfort conducive to the creative process and the creator's well-being. How can we support creators and reduce the negative emotional burden that too often accompanies the projection mapping production process?

5.5. New residence arrangements

5.5.1. *Limitations and contributions of this type of survey*

To try to understand what are the meaningful constructs of projection mapping residents in natural creative situations, we used the REMIND method, which uses the recording of the subjective video trace of the activity of the creators surveyed to stimulate their subsequent recollection. It allowed us to identify significant segments of their experience based on the body, cognitive and emotional dimensions experienced in the course of their activity. Thus, we were able to rebuild the dynamic of creation. Of course, the methodological temporal coring of this study only reflects the activity during the times when the creators are equipped. By using the REMIND method, the subjective perspective reproduces the creator-environment coupling in the same way as this coupling had previously given rise to experience. When interviews are well conducted and their reflective conscience is not mobilized, creators do not watch a video recording, they relive their own story first and foremost. The wearing of the eye-tracker does not seem to introduce any significant bias, especially since the creators have had the opportunity to get used to it. When creators wore glasses, we equipped them with a mini tactical camera. The point of view of subjective video recording is similar to that of an eye-tracker.

5.5.2. *Towards a design of space and experience*

According to Marie-Christine Bordeaux (2014): "You cannot express your freedom and creativity in a defined and imposed space." The adequacy of the workspace with the primary expectations of creators seems much more important than we assume. In addition, the time it takes to adapt to the work environment influences the creator's work process, gestures, cognitive and emotional state. It can be both inspiring and rewarding as well as frustrating and uncomfortable. Indeed, software imposes formal frameworks that strongly constrain the expression of creators. Being a projection mapping creator in this case means being calculating, that is setting up as a technician, as a craftsman who produces something while respecting the technical constraints, the strategies pre-established by others. As Ignace Meyerson says: "The man in front of the tool can be master or cog; he can feel more or less dependent; he can participate more or less in a diverse way in the machine. The machine must certainly help man to accomplish a task, but he must 'undergo' an apprenticeship to benefit from this assistance" (Meyerson 1948 in Rabardel 1995, p. 31). If he has the expertise, the creator can consider hijacking software and inscribe his uniqueness in it. However, in this technical and digital world of programming with highly constrained residence times, is there enough space for improvization and exploration for novice creators? In their view, the young residents we followed did not seem to have succeeded in diverting formal frameworks and taking advantage of constraints in a satisfactory way. A team of expert supervisors seems necessary to install the creator in an environment that makes sense to him, that suits him and then assists him in the complexity of projection mapping in order to overcome technical pitfalls and remain focused on creative intentions.

5.5.3. *The creator profession*

The process of setting up a projection mapping piece is as slow and tedious as the preparation of a wooden panel in the 15th Century. Whether it is to handle a paintbrush, a pencil, a brush, a scraper or to master the language of a software, creation requires know-how. In the case of digital images, creators "are led to use abstract modes of thought to control formalized operations in an environment of codes and messages" (Lévy 1987, p. 11). The projection mapping creator must know how to master a computer language to guide his creative gesture in digital images. Most often, creators succeed in seizing computer technical tools by trial and error, improvising or end up suffering from these problematic situations. It is rather the latter situation that has been observed among residents. The analysis of their emotional states reveals more suffering than discomfort. Residents find the creative process slow, boring, repetitive. The artistic dimension appears at the end of the process, when all the steps are completed and the final movement is born. This feeling reminds us that the creative activity is above all an effort, an activity that requires a sustained effort to produce a new idea. The "artistic profession" or "artistic work" as it is experienced on a daily basis is far removed from the image of Épinal as the "cursed artist" of the 19th Century or from a majestic activity.

5.6. Prospects for the future

This study focused on debutant creators in residence. It would probably be fruitful to apply the same method to experienced practitioners, trained in many tools and compare their creative dynamics. For example, does an experienced artist-engineer experience so many pitfalls with his tools? How does he consider his production space? Moreover, could projection mapping evolve and gain in maturity based on a particular attention paid to the tools and needs of creators? Would the fact of being able to work in a space that corresponds to the creator's needs and with the devices of his choice renew projection mapping creations? Would it open up forms of expression not previously exploited by the medium? We formulate the hypothesis that the material environment and technical support could favor the creative evolution of projection mapping. A projection mapping that sometimes tries to match the modalities of professional film production while mobilizing human and financial resources that are more a matter of amateur cinema.

As for the words of creators gathered during the creative process, can we imagine other purposes to this word than that of this study? Understanding the process of production or even creation, capturing part of the intimate thought of creators could add a little-known dimension to material works and their interpretation. The material of the work is part of the heritage and deserves to be preserved and shown to all as much as the intangible heritage it also contains: the ideas, emotions, etc., that have accompanied the creative process. We can now access these testimonies with a method like REMIND. Of course, it is not a question of generalizing the use of such a method, but of making it known in the same way as interviews or recordings with creators. Could the exhibitions enrich the display of the works of their authors' testimonies? Audiences and art historians would then have access to the creator's journey from the genesis of an idea to the finished work. Especially since there are several entries into art: certainly there is the view, but also the sensitivity of ideas and concepts that emerge from an approach. Conceptual art may have a new mechanism for mediation by creating a dialogue between the process of the work, the object that symbolizes it and the audiences that observe it. In the audiovisual field, it would be interesting to make documentaries from the subjective perspective and the revival of artists. And if verbalizations retrace part of reality, part of an experience, is it possible to fictionalize them to unfold an artist's life trajectory? Finally, without seeking to produce a new work, depositing the recording of creators' own worlds on an open archive could build and form a historical trace of a given time.

5.7. Increased attention to the place of creators in digital arts

When we organize artist residencies, particular attention should be paid to the constraints and their consequences on the creative process. Art is experienced by our creators above all as a practice. They produce works with specific means, materials, skills and operations that require know-how and supervision. Too short an execution time, insufficient mastery of the tools and isolation mark the experience of creators much more than they exalt creation. And when these factors add up, the experience can become anxiety-provoking and inhibit the creative process, which also opens up avenues for improving the conditions for creation in residence. Persisting, even unconsciously, in considering the artist as a creative genius does not serve his cause. Far from the stereotypical image of the bohemian and idle artist, artistic work is organized in time and space. The doxa of digital technology accessible to all, which facilitates any activity, leads us to deny the difficulties of novice creators. Listening to creators, their learning sequences of the tools do not seem to have contributed significantly to their production: they have experienced them as obstacles, hurdles to the creative process. It also means that not all constraints serve the creation well.

Does the mere completion of a production mean that its author has benefited from it? It is utopian, according to Menger, to think today that the activity of artistic creation is extra-economic (Menger 2002). The idea that artistic work is ideal, desirable and that, unlike salaried work, it is not alienated, it systematically allows the subject to accomplish himself while he deploys all his capacities and that his action is not transformed into a means to obtain a gain no longer seems topical. On the contrary, the artist, in producing, does indeed fit into the social and economic sphere. Today, these artists are confronted with the different struggles imposed by this system. From an economic point of view, these struggles can be explained by the level of training, knowledge and experience specific to each artist. According to Menger, the presence of inequalities of success in the art world, so little different from other inequalities in work, are not questioned. For him, we would be in an area of talent development. And the talent selected benefits from a team of professionals of comparable talent in the other professions necessary for its production: the very good director of the moment, for example, can call on the services of a very good scriptwriter, a very good editor and a very good director of photography. The permanent spectacle of the lives of certain artists and their success contributes to maintaining these illusions, never evoking the ordinary social condition of artistic workers. Yet creators often move from one project to another with great enthusiasm, even though the contracts are short-term, of little intellectual, material or financial benefit to them. The artist is adaptable, flexible, versatile, active, autonomous and capable of change (Boltanski and Chiapello 2011). This idealized image of the contemporary artist inspires many other fields than art, while at the same time suffering from the global evolution of the precarious labor market. In short, the young artist in residence and without notoriety does not benefit much today from a creative environment that takes into account in a balanced way his actual work, his remuneration, his continuous training in new tools and, more generally, his well-being.

5.8. Acknowledgements

In 2017, the Hauts-de-France region highlighted the importance of the creative industries by including them in the Regional Economic Development, Innovation and Internationalization Scheme. We would like to thank the association Rencontres Audiovisuelles (Lille), the DeVisu laboratory of Polytechnic University of Hauts-de-France and Arenberg Creative Mine with the support of the European Union, the ERDF programme, the Hauts-de-France region and the urban community of La Porte du Hainaut for having organized the first edition of a projection mapping festival and enabled this research.

5.9. References

Aubert, O., Prié, Y. (2005). Advene: active reading through hypervideo. In *Proceedings of ACM Hypertext'05*. Salzburg, Austria.

Bédard, P. (2014). *L'art en pratique. Ethos, condition et statut social des artistes en arts visuels au Québec et en Belgique francophone*. Université du Québec à Montréal/Université libre de Bruxelles, Montreal/Brussels.

Boltanski, L., Chiapello, E. (2011). *Le nouvel esprit du capitalisme*. Gallimard, Paris.

Bordeaux, M.-C. (2014). Pour un réexamen de la notion d'usage: la dimension culturelle de l'expérience. *Lendemains – Études allemandes comparées sur la France*, 154/155, 76–100.

Grésillon, A. (2016). *Éléments de critique génétique: lire les manuscrits modernes*. CNRS, Paris.

Hockney, D. (2001). *Secret Knowledge: Rediscovering the Lost Techniques of the Old Masters*. Studio, London.

Lévy, P. (1987). *La machine univers: création, cognition et culture informatique*. La Découverte, Paris.

Menger, P.-M. (2002). *Portrait de l'artiste en travailleur. Métamorphoses du capitalisme*. Le Seuil, Paris.

Menger, P.-M. (2009). *Le travail créateur. S'accomplir dans l'incertain*. Le Seuil, Paris.

Pignier, N. (2012). Le plaisir de l'interaction entre l'usager et les objets TIC numériques. *Interfaces numériques*, 1, 123–152.

Rabardel, P. (1995). *Les hommes et les technologies; approche cognitive des instruments contemporains*. Armand Colin, Paris.

Rautureau, M. (2014). Video Mapping: pratiques contemporaines d'un nouvel art élargi en France. Université Paris III, Paris.

Schmitt, D., Aubert, O. (2016). REMIND, une méthode pour comprendre la microdynamique de l'expérience des visiteurs de musées. *Revue des interactions humaines médiatisées*, 17(2), 43–70.

Theureau, J. (2006). *Le cours d'action: méthode développée*. Octarès, Toulouse.

Varela, F., Thompson, E., Rosch, E. (1991). *The Embodied Mind: Cognitive science and human experience*. MIT Press, Cambridge.

Vermersch, P. (1994). *L'entretien d'explicitation*. Sociales françaises, Paris.

6

Projection Mapping and Automatic Calibration: Beyond a Technique

6.1. Introduction

Today, tracking technologies make it possible to track objects moving on stage by adapting the projected video content, while operating automatic calibration techniques (calibration of the video projectors). These technologies culminate in a new approach in terms of application: projection mapping (or video mapping) is not just a technique, but a media format, a new medium for spatialized entertainment. This type of projection is used when projecting a complex surface onto an object, a façade, a 3D or relief structure. The difference between projection mapping and a standard projection is anamorphosis: the projection screen is not flat, and the projection must therefore be deformed accordingly; this process consists in correcting distortions on the different parts of the image by means of different deformations and geometric curves.

6.2. Towards a new projection dynamic

Interactive projection mapping is when the projection is modified in real-time following a particular event. This can be the effect of an external device: a mobile phone connected to the server on which the projection mapping software operates, a tablet, or even an infrared camera. On/off switches can also be considered, but this is more of a reactive than an interactive system.

Chapter written by Sofia KOURKOULAKOU.

Tracking is a technique that allows us to follow, by dynamic projection methods, a moving object: it means that the machine interprets and deducts its position in the represented and real 3D space. The machine also anticipates the next movements of the object so that the content projected on the object is automatically deformed to be calibrated in real-time during its movement.

6.3. Automatic calibration

The automatic calibration of projectors is a method of automatically calculating their anamorphosis. Until now, it has been possible to perform manual calibrations, requiring technical teams to deform the key points of the video content directly on the projected surface themselves, resulting in a significant amount of work time.

To meet a growing need, the researchers developed computer algorithms to observe the projection on the surface and estimate the deformation required to display geometrically and photometrically correct images. This is done using a camera/projector system called procams. Several engineers have addressed the subject of automatic projector calibration as a research and development topic. Applications can be classified into two categories (Grundhöfer and Iwai 2018):

1) those who wish to project without geometric distortion of the content onto a complex geometry (geometric calibration);

2) those that are able to flexibly control the appearance of the projection colors (colorimetric calibration).

To achieve such objectives, several calibration steps must be performed, which can be described as follows:

1) geometric calibration steps allowing the acquisition of the exact shape of the projection surface, related to the internal and external parameters of the projectors and cameras used;

2) photometric calibration tasks with the estimation of the internal color processing of the input and output devices used, as well as the reflectance properties; reflection factor of the surfaces to be projected (Grundhöfer and Iwai 2018).

6.4. Automatic geometric calibration

6.4.1. *Procams methods*

Most projector calibration methods usually start with one or more cameras, which are initially pre-calibrated or uncalibrated. Cameras are used to capture a series of structured projections (light patterns) to generate the correspondences between the projector and the camera pixels, all with an accuracy in the pixel range or even lower than the pixel (sub-pixel). In combination with a multi-camera geometric calibration procedure, which estimates their optical properties and overall orientations relative to each other, the surface geometry can be reconstructed.

6.4.2. *Zhang method (Zhang 1998, 1999)*

This method is one of the first really successful methods, so it is always a reference in relation to the others.

Zhang's technique involves a pinhole camera and projectors. Calibration is performed by pointing the uncalibrated camera at the surface where the image is projected. The camera records the image obtained from its point of view, which then compares it with the original image and the intrinsic and extrinsic geometric parameters of the projector. The intrinsic parameters taken into account are focal length, pixel size and skew factor. The extrinsic parameters are the translation and referential rotation with respect to the camera's *world* reference frame. This method assumes that the camera does not have any distortion.

6.5. Projector calibration using one or more pre-calibrated cameras

With at least two calibrated cameras, a projector calibration can be performed using structured light patterns to generate matches between all cameras and projectors. As the cameras are already calibrated, the maps can be used to triangulate a cloud of points on the surface, then use this information to record the projectors on the surface using the 3D-2D point-to-pixel maps to estimate the projection matrix, using the direct linear transformation (DLT) approach and an additional non-linear method. To ensure photometric uniformity, several other color and intensity models are projected and analyzed to estimate the processing of the internal color of the device as well as the surface light. In combination with the geometric information, the overlapping of projectors is gently processed to finally generate a photometrically homogeneous projected image. This method offers an optimization of the calibration of intrinsic and extrinsic parameters.

The calibration of the intrinsic parameters can be performed in several ways. Often, a flat matrix representing a checkerboard is used: the camera captures the projected checkerboard image in order to find the reference points and calculate the focal distortion. If the geometry of the projected surface is known, calibration can be done manually without the need for a camera. This principle is performed by manually searching for a correspondence between the points marked on the surface, and the points on the same 3D object whose geometry is known. However, this method is less reliable because it incorporates a human error factor. Based on the Zhang method, the calibration process is programmed according to the following steps (Anwar *et al.* 2012; Moreno and Taubin 2012):

1) images of an object in regular form (such as a checkboard pattern) are captured at different positions and orientations;

2) the characteristic points of the image are detected using a function and stored as "image points";

3) the coordinates of these characteristic points in the global coordinates of the system are stored as "object points";

4) the two matrices are compared to find the distortion and intrinsic parameters of the camera;

5) all the extrinsic parameters of the camera are then determined using the intrinsic parameters. The projectors are geometrically calibrated to generate a coherent projection image.

6.5.1. *Fringe Pattern/Structured Light DMD (Digital Micromirror Device)*

Other methods propose the calculation of the image captured by the projector using horizontal and vertical images of structured light: fringe light system (Li *et al.* 2008). Here, the projector is treated like a camera, to unify the calibration procedure using structured light and the stereovision system.

This method consists of establishing an extremely close match between the camera pixels and the projector pixels, and to generate DMD (digital micromirror device DMD image sets) image sets for calibration. A phase shift method is then used to align the DMD images with the CCD (Charge Coupled Device images) (Fiala 2005; Li *et al.* 2008).

6.5.1.1. *Planar Homographies*

Gao *et al.* (2008) propose an automatic calibration using the homography method on a plan. *(Planar) Homography* refers to a 3x3 matrix that guesses a homogeneous linear transformation from one plane to another in the projected space.

6.5.1.2. *Reprojection error*

Several researchers and laboratories have been trying for years to find a more precise method to compensate for the error calculated on the reprojected image. This geometric error is the image distance between a projected point and a measured point.

6.6. Automatic calibration applied

Today, autocalibration tools are being developed as part of applied research. Thus, projector manufacturers offer geometric and colorimetric autocalibration with specialized software. Calibrize, open source software, helps to correct the colorimetric correction of your devices (displays). Projection mapping software or media servers offer automatic calibration as an integrated solution in their operation. In the United States, several software programs offer this solution, in order to ensure large monumental projections at the professional level. The Scalable software manages the automatic calibration and is specialized in display technologies. In Europe, Avolites manages calibration using light patterns. DomeProjection has developed a toolbox for versatile automatic calibration. Stumpfl, a German software, offers automatic calibration among other sophisticated Computer Vision features. Another German software, Vioso, offers automatic calibration and is specialized in dome calibration. The D3 software uses OmniCal technology for automatic projector calibration. In Sweden, the Watchout software offers semi-automatic calibration and has also integrated the DomeProjection system for autocalibration.

6.7. Automatic calibration in France

Modulo Pi, a French company based in Paris, presented the Kinetic Designer to ISE (Integrated Systems Europe) in February 2017. More than a media server, it is a *hardware-based software*, which follows technicians and content creators through all stages of the installation of projection mapping systems. Toolbox designed with ergonomics in mind, Kinetic Designer can therefore manage not only manual calibration (*2D warping*), but also automatic and semi-automatic calibration of projectors. With an integrated video projector library, you can choose the focal lengths you will use and simulate the projection in your 3D engine by placing virtual projectors, study the brightness and pixel density of the projection in order to optimize the installation beforehand. Kinetic Designer makes dynamic mapping or object tracking possible.

Figure 6.1. *Launch of Modulo Kinetic Designer, at ISE (photo credit: Holymage, 2017). For a color version of the figures in this book see, www.iste.co.uk/schmitt/image.zip*

6.8. Conclusion

For more than 10 years, automatic calibration methods have been evolving at an accelerated pace, perfecting video projection applications more and more. Mapping is a media experience at the border between spatial augmented reality and wonder; we can expect dynamic projection mapping and automatic projector calibration to grow quickly in the coming years.

6.9. References

Anwar, H., Din, H., Park, K. (2012). Projector calibration for 3D scanning using virtual target images. *International Journal of Precision Engineering and Manufacturing*, 13(1), 125–131.

Fiala, M. (2005). Automatic Projector Calibration Using Self-Identifying Patterns. In *Proceedings/CVPR, IEEE Computer Society Conference on Computer Vision and Pattern Recognition*.

Gao, W., Wang, L., Hu, Z.-Y. (2008). Flexible calibration of a portable structured light system through surface plane. *Acta Automatica Sinica*, 34(11), 1358–1362.

Grundhöfer, A., Iwai, D. (2018). Recent Advances in Projection Mapping Algorithms, Hardware and Applications. *STAR*, 37(2).

Li, Z., Shi, Y., Wang, C., Wang, Y. (2008). Accurate calibration method for a structured light system. *Optical Engineering*, 47(5).

Moreno, D., Taubin, G. (2012). Simple, Accurate and Robust Projector-Camera Calibration. In *Proceedings of the 2012 Second International Conference on 3D Imaging, Modeling, Processing, Visualization & Transmission*. Zurich, Switzerland.

Zhang, Z. (1998). A Flexible New Technique for Camera Calibration. Technical Report, Microsoft Research, Redmond.

Zhang, Z. (1999). Flexible camera calibration by viewing a plane from unknown orientations. In *Proceedings of the Seventh IEEE International Conference on Computer Vision, INSPEC*.

7

Projection Mapping Gaming

7.1. Introduction

A characteristic of video games is that they offer an analog or digital display as an outgoing interface. The screen usually plays this role. We find the idea of associating screens with games as early as the early 1950s if we refer, for example, to *Oxo*, a tic-tac-toe game programmed on the Cambridge University *EDSAC* computer by the English computer scientist Alexander Shafto "Sandy" Douglas (Djaouti 2019, p. 11). It should also be noted that in the late 1940s, Thomas Toliver Goldsmith Jr. and Estle Ray Mann filed a patent for the *Ray Tube Amusement Device Cathode* (Djaouti 2019, p. 3), a game where you must succeed in firing a missile at a target., all played on a CRT screen. However, as no trace of the device has been found to date, it is not yet known whether it is just a paper invention or whether a prototype has been developed.

Thus, this link between screen and video game has long been the norm according to Mark J.P. Wolf to define the video game (Wolf 2008, pp. 3–8). The very nature of screen technology was a differentiating feature. This made it possible, for example, to distinguish between video games and computer games. This positioning thus explains disagreements between video game experts when it comes to defining the first video game in history. Since *Oxo* was played on an oscilloscope, it is not really a video game in the view of some specialists. So, for them, it is better to wait until 1962 and the advent of *SpaceWar!* This game, developed at MIT by Martin Graetz, Wayne Wiitanen and Steve Russell, is played on a CRT monitor

Chapter written by Julian ALVAREZ.

(Wolf 2008, p. 40). Wolf then explained that the definitions of video games have evolved around the screen to take into account other technologies such as liquid crystal displays (LCDs) in *Game & Watch* games launched by Nintendo in the early 1980s. Over time, it was finally the display of pixels, regardless of the process, that seemed to prevail (Wolf 2008, pp. 3–8).

In this dynamic, if we explore the history of video games as an industry since 1972, when the game Pong (Atari 1972) was introduced and the *Magnavox Odyssey* home console (Magnavox 1972) was released on the market until today, several types of displays can be referenced, as shown in Figure 7.1.

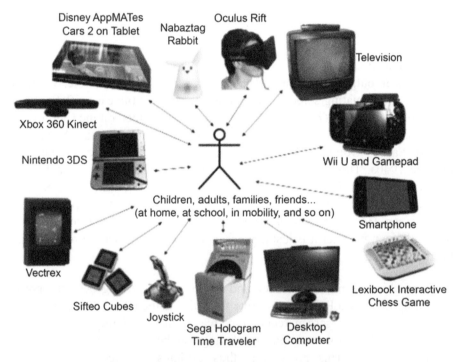

Figure 7.1. *Different types of displays associated with video games (source: Alvarez et al. 2014). For a color version of the figures in this book see, www.iste.co.uk/schmitt/image.zip*

There is a plethora of display technologies ranging from analogue to digital technologies, including monitors, televisions, arcade terminals, virtual reality headsets, tablets, smartphones and even robots equipped with light-emitting diodes.

This plurality of screens presents us with a wealth of outgoing interfaces that definitively bury the debates between oscilloscope and monitor.

However, in the inventory shown in Figure 7.1, it should be noted that the video projector is missing. However, while on-screen display is a very widespread technology in video games, it is also possible to project the virtual environment of a digital game through video projection. This corresponds well to the display of pixels. In the register of video games calling for video projection, we can identify as early as 1988 an Italian game console, called the *Mega Projector Video Game*, also called *Light games*. The technology is based on the old-fashioned slide projector, where the user is offered the option of inserting transparent liquid crystal game cartridges[1]. Under the effect of a lamp and a magnifying lens, the virtual environment of the game is projected on a wall with a diagonal of 85 centimeters according to the mention made on the box of the device (Figure 7.2).

Figure 7.2. *Game console: Projector Mega Video Game/Light Games (source: Playtime 1988)*

---

1 A video produced by Pocket Legion shows how the *Projector Mega Video Game/Light Games* game console works: www.youtube.com/watch?v=jNyz3eaZK40 (accessed January 10, 2019).

Nowadays, given the accessibility of video projectors and pico-video projectors in terms of price, it is possible for a growing proportion of the general public to connect game consoles or computers to such outgoing interfaces.

This gives the possibility to offer game experiences with larger audio-visual displays. However, large LED and Oled technology displays are becoming more popular and therefore offer similar experiences. As a result, the use of video projection must now be justified *through* other experiments. In this respect, the projection mapping trail is a path to explore. But is this an area that can really be a privilege for video projection? This is what we will check in this chapter by first defining what we mean by projection mapping game. With this in mind, we will then identify the main types of video mapping games and specify the associated technologies. We will thus see if the projection mapping game linked to rear projection presents a technological lock with regard to the screens. At the same time, this will lead us to question whether video projection really makes it possible to distinguish projection mapping from video games.

7.2. Specifying the scope of the projection mapping game

In the field of video games, mapping can mean different things. To begin with, this can correspond to the action of producing maps to define the different game levels of a game. Mapping refers to the level design, i.e. the process of designing and editing a game level.

Mapping in the field of 3D computer graphics can also mean applying a texture to an object. In the jargon, we talk about "mapping" a texture. For a real-time 3D video game, the way to map textures is strategic. Thus, not applying texture to areas inaccessible by the camera helps to optimize the machine's performance and make the game more fluid on the display.

In this chapter, the term mapping refers to the "video mapping" that underlies the idea of projecting an image with an illusionist aim on an object or surface, for example the façade of a monument or a building. Thus, based on the projection mapping approach, we can clarify what we mean by "projection mapping game": it is a digital game whose virtual environment is video-projected onto a given surface with an illusionist aim with the possibility of interaction. If the notion of "game" by nature is complex to define because it is subjective, we cannot establish a property around this notion except that the game designers define it as such and that the target audience recognizes it as such.

On the other hand, we can at least use three objective criteria to define the properties of a projection mapping game: 1) video projection; 2) illusionism; 3) interactivity.

Specifying these three criteria makes it possible to specify the devices or works that we can now remove from the scope of the projection mapping game. The "video-projection" criterion excludes all video games played only on screen. This also applies to games that combine objects connected to interactive tablets that can give certain titles an illusionist dimension. In this register, there is, for example, *Safari in a balloon* by Éditions volumiques (2014)[2], which combines an interactive tablet and a paper hot air balloon. If the shadow games associated with zooming on the landscape in plane mode on the tablet gives the illusion of a movement of the balloon, we miss the video projection. For its part, the "illusionist" criterion excludes devices such as the *Mega Projector Video Game* console (Figure 7.2) that only target video projection on a flat surface. Finally, the "interactive" criterion excludes non-interactive shows, including works with a video game theme such as Pavel Novák[3]'s *8 Bit Invader*, a 2011 work that pays tribute to a series of video game titles from the 1980s, or the show proposed by publisher Blizzard in 2015 in London to promote the launch of the video game *Heroes of the Storm*[4], or Tonner Vi's works that produce projection mapping on game consoles[5].

Now that we have clarified what we mean by projection mapping game, we can conduct our state of the art by exploring the first devices to trace back to those that can be identified in the late 2010.

7.3. The indoor projection mapping game

We shall begin our exploration with digital game devices associated with indoor projection mapping. In this register, we can start by identifying the

---

[2] https://vimeo.com/118038081 (accessed January 10, 2019).
[3] https://laughingsquid.com/8-bit-invader-video-game-projection-mapping-on-a-building/ (accessed January 10, 2019).
[4] https://youtu.be/bygoTpNF86A (accessed January 10, 2019).
[5] www.vice.com/en_uk/article/bmdk8d/3d-projection-mapping-with-your-favorite-nintendo-games (accessed January 10, 2019).

Sega Hologram Time Traveler (Sega 1991) terminal, which since 1991 has offered a "holographic" display. This illusionist effect is obtained with a set of mirrors[6]. This only works on a small scale in a dedicated space within the device.

This device, which is quite unique in the history of video games, could undoubtedly be considered as one of the first projection mapping games. However, since video projection takes place here within the system itself, it could potentially lead some specialists to exclude the *Sega Hologram Time Traveler* from projection mapping games in the same way that some specialists exclude *Oxo* from the video game family, as discussed in the introduction. Without entering into such a debate, we will remain cautious by continuing our investigations with devices proposing extrinsic video projections.

In this respect, we can start by identifying an artistic installation entitled *3D Pac-Man in* 2013. Proposed by the Savannah Art Museum in Georgia (USA), it invites the experimenter to direct the Pac-Man with a controller and move it on the different walls and ceilings of a dedicated room[7].

Still in the artistic cultural field, it is possible to identify the Augmented Pinball device (464 Medialab 2014). It is a wooden pinball machine where the balls are video projected[8]. It should be noted that this device pays tribute to Kandinsky in the context of a temporary exhibition held in 2014 at Seoul City Hall in South Korea. If this installation has been custom-made in Kandinsky's style, it is also possible to use works to play with it. For example, nine students proposed this in 2018 at the Palais des Beaux-Arts in Lille as part of a four-day creative challenge linked to the Video Mapping Festival #1 and the European Euranim project. Their creation, entitled *Pieces of peace*, invites participants to play with the *Pax statue* of Edgar-Henri Boutry (1938). Players interact with their mobile phones to find the right parts and colors to apply to the statue[9].

---

6 The following video shows how the Sega Hologram Time Traveler terminal works: www.youtube.com/watch?v=Y-SuK5-cKk0 (accessed January 10, 2019).
7 www.clementshimizu.com/3d-pac-man/ (accessed January 30, 2019).
8 https://vimeo.com/118664449 (accessed January 30, 2019).
9 www.geraldinekwik.com/project/video-mapping-festival-1-opera-de-lille-ouverture-4-2-2-2/ (accessed January 30, 2019).

In the case of works for museums, the offer to industrialize projection mapping games for private individuals in the domestic context is also envisaged. Microsoft, which produces the *Xbox One* game console (Microsoft 2013), announced a project in 2013 called *IllumiRoom* (Microsoft Research 2013) to increase the visual experience around the game screen[10]. In concrete terms, it is a video projection with animations that are linked to the virtual environment of the video game used. Among the video-projected animations, some are clearly part of an illusionist approach, such as game elements that seem to come out of the television set or effects that seem to distort and make the furniture shake around the screen. In 2014, the project is renamed *RoomAlive*[11] and proposes to free itself from the screen to transform a room into a real cellar (immersive room). Innovative gameplay around this technology is also presented[12], such as catching or neutralizing virtual characters on the different walls of the room. Although the approach is clearly in line with the projection mapping game industry, this technology is still not available at the end of 2010. Perhaps the strong development of virtual reality since 2013 is competing with this technology? This suggests that screen technology might propose some form of head-on competition at this stage.

In another area, Hiroshi Ishii, of the Tangible Media Group and Ars Electronica Center, launched the *PingPongPlus* device in 2001, which takes the form of a traditional table tennis table on which a 3D image of water with fish is video projected. With each impact of the ball, water circles materialize and attract the school of fish. This is more of a toy as there is no objective set by the device to win or lose.

More recently, since 2016, there has been a similar concept called *Table Tennis Trainer*, a prototype developed by Thomas Mayer to[13] allow players to train and study their performance. With the variables studied by this prototype, it is possible to consider objectives with scores to be achieved. This brings us back to the possibility of offering a game with objectives.

---

10 www.youtube.com/watch?v=m8xrH-Z2cVE (accessed January 30, 2019).
11 www.gamehope.com/news/30828-illumiroom-devient-roomalive.html (accessed January 30, 2019).
12 www.youtube.com/watch?v=ILb5ExBzHqw (accessed January 30, 2019).
13 www.ulyces.co/news/cette-table-de-ping-pong-interactive-va-faire-de-vous-des-pros/(consulted on January 30, 2019).

This is precisely what the English company Bounce has been offering us since 2017, which has been broadcasting the *Wonderball* system (Bounce 2017). It is a ping pong table on which a set of mini-games is projected[14]. Public places such as bars are targeted by this type of device.

Still in the sports field, we also list the Augmented Climbing Wall developed in Finland by the University of Aalto and Valo Motion since 2013[15]. It is interesting to note, in the field of gaming, the adaptation of the title *Pong* (Atari 1972) on this device. Called *Climball*, this game is played by two people, ideally against one another, on the climbing wall to send the ball back to the opposing side. *Climball* means climbing up and down as fast as possible[16].

The Game Research department of the University of Aalto is working on the implementation of different games around the climbing wall in augmented reality such as: *Astromania*, which involves returning meteors to space before they hit the Earth; *Whack-a-Bat*, which involves hunting bats hidden in the walls; and finally *Sparks*, which challenges the player to climb by avoiding moving white tracks representing electrified wires[17]. All these different titles are presented as allowing participants to train for climbing, and thus exercise. This allows us to assimilate such games to Serious Games.

The field of sport seems to be a fertile ground for identifying devices that offer to display information on the playing field. Basketball is for example concerned by such an initiative through the *Nike Rise system* (Nike 2014)[18] or *Nike Rise 2.0* (Nike 2016)[19]. However, these are LED-based technologies. The field is a giant screen. Here again, we find the frontal competition between the screens and the video projection mentioned in the introduction.

This observation invites us to change categories and switch to outdoor projection mapping games involving large displays. This change of category is the continuation of our exploration and allows us to verify if video projection can stand out from screen technologies.

---

14 https://vimeo.com/236600735 (accessed January 30, 2019).
15 www.youtube.com/watch?v=vuLgcN5GosE (accessed January 30, 2019).
16 www.youtube.com/watch?v=cuhlknicMEA (accessed January 30, 2019).
17 www.vrfitnessinsider.com/climb-new-heights-valo-motions-augmented-climbing-wall/ (accessed January 30, 2019).
18 www.youtube.com/watch?v=u2YhDQtncK8 (accessed January 30, 2019).
19 www.youtube.com/watch?v=nGuOA_EJ8qk (accessed January 30, 2019).

7.4. The outdoor projection mapping game

From the *Augmented Climbing* Wall to the display on building facades, there is a change of scale that the projection mapping game is able to cross. There are, in fact, several achievements in this area. In chronological order, we can start by identifying the game *Tetris* (AcademySoft CCAS USSR Moscow 1984) which has been the subject of several video game mapping adaptations. In this respect, since 2007, there has been the possibility of playing the Russian title on the facade of the *Brown's Tech House* building in the city of Providence (USA). This approach involves turning on and off the hundred or so windows on the building's facade to simulate the various quadraminos of the game[20]. This principle was taken up by MIT in 2012 with the *Green Building*. This time, the lights are colored[21]. In 2016, a more sophisticated version based on LEDs is offered to the general public on the facade of Tel Aviv City Hall in Israel[22]. However, we cannot associate these games with projection mapping as video projection is not appropriate. Once again, screen technology seems to be competing with video projection. To find a projection mapping game dedicated to *Tetris*, it is advisable to turn to the Mexican communication company Koi Comunicación, which launched the game *Espacio Tetris* in 2014. The *Tetris* game is projected on the facade of the *Espacio Las Américas*[23] shopping mall. If we count video projection, the "illusionist" criterion is not really identified because the game is played on a large flat surface like a giant screen.

To identify a projection mapping game that makes more use of the architecture of the facade, we identify the *Pac-Man* game (Namco 1980) which, in 2013, was video-projected on a building in Jerusalem[24]. In 2016, MediaCutlet offered projection mapping games based on the title *Pac-Man* to showcase its know-how in a building patio[25]. The different edges of the façade represent in this case the different paths that the Pacman can take.

It is interesting to note that, as of 2012, the game *Snake* has also been adapted for projection mapping. Named *Snake the Planet* (Mpulabs 2012),

---

20 www.cnet.com/news/brown-students-create-massive-tetris-game-on-building/ (accessed January 30, 2019).
21 http://hacks.mit.edu/by_year/2012/tetris/ (accessed January 30, 2019).
22 www.youtube.com/watch?v=bgch5VkwX3M (accessed January 30, 2019).
23 www.youtube.com/watch?v=Z8kSOqC5FMMM (accessed January 30, 2019).
24 www.youtube.com/watch?v=teOSyK9aAHI (accessed January 30, 2019).
25 www.youtube.com/watch?v=4uVNtOoY65g (accessed January 30, 2019).

this game is associated with MPU Urban Canvas technology[26], which automatically generates the different levels by taking into account the various elements of a building's façade: windows, doors, etc. Players can also come to position themselves in front of the building's façade to generate additional obstacles. This technology makes it possible to facilitate the *level design* (creation of levels) for projection mapping games.

In addition to creations based on classic video game history, it is interesting to note original creations. In this context, the game *Trouve BoB* (GPG/Champagne Club Sandwich 2013) was proposed in 2013 at the University of UQAM in Canada[27]. This installation, which takes a UQAM building as its display, offers the player the opportunity to find the BoB character in a very colorful setting before time runs out.

In France, since 2014, the Play-Foul collective has been offering the projection mapping game entitled *RGB Racers*[28]. In concrete terms, it involves offering three players the opportunity to compete in a race of small spaceships that navigate in colorful and pixelated settings reminiscent of 8-bit games. In this installation, it is interesting to note the possibility for players to interact with steering wheels instead of controllers or smartphones (Figure 7.3).

Figure 7.3. *RBG Racers (source: Play-Foul 2014)*

---

26 vimeo.com/37637793 (accessed January 30, 2019).
27 www.youtube.com/watch?v=Q5cOdzuPEPo (accessed January 30, 2019).
28 www.play-fool.net/web_data/page6/fold7581537/RGB-Doc-2018_A_-fichetech.pdf or www.play-fool.net/rgb-racers (accessed January 30, 2019).

In 2015, the projection mapping game *Giant Play* was offered in Mons (Belgium). It allows one to three players to capture dragons that appear at the windows of a building facade before they escape. The goal is to get the best possible score.

It should also be noted that the number of players may be even higher for some projection mapping games. Thus, the game *Sciences Poule*[29], produced by a group of 13 students in 2018 as part of the Video Mapping Festival #1 and the European Euranim project, offers several dozen participants the opportunity to interact with their smartphones. The principle is to vote for the different choices that the avatar must make. This voting system probably reflects the fact that this game was screened at Sciences Po Lille.

7.5. Conclusion

Starting from the idea that video games could free themselves from screens by adopting video projection, there appeared a competition between Led/Oled and video projection technologies. Thus, as the screens become larger and larger, video projection must be distinguished by the "illusionist" criterion. Indeed, to be in the presence of a projection mapping game, it is necessary to identify three criteria: video projection, an illusionist aspect and interactivity.

Once these criteria had been identified, an exploration of video game devices in line with projection mapping was carried out. Two sub-categories of devices were explored: indoor and outdoor.

The devices identified in these two categories made it possible to identify the instances that LED/Oled technologies were present. For the interior, there is, for example, the *Nike Rise basketball* court. For the exterior, we identify the *Tetris of* Tel Aviv City Hall. The projection mapping game based on video projection therefore seems to be overtaken by screen technology on all fields.

This observation ultimately raises questions about the relevance of retaining the "video projection" criterion to define projection mapping games. Indeed, if LED/Oled technologies are able to allow illusionist effects of small and large scales by offering interactivity, is there still a prerogative

---

29 https://vimeo.com/263312277 (accessed January 30, 2019).

of video projection? Such a question puts the debate of specialists into perspective when they try to determine if *Oxo* was really a video game because it works on an oscilloscope. Returning to our exploration of projection mapping games, does the outgoing interface, screen or projector constitute a criterion that differentiates it from digital games? If the use for the player remains ultimately similar, can we really distinguish projection mapping from video games?

7.6. References

Alvarez, J. (2007). Du jeu vidéo au serious game, approches culturelle, pragmatique et formelle. Information and Communication Sciences PhD thesis, Université de Toulouse, Toulouse.

Alvarez, J., Haudegond, S., Havrez, C., Kolski, C., Lebrun, Y., Lepreux, S., Libessart, A. (2014). From screens to devices and tangible objects: a framework applied to Serious Games characterization. In *16th International Conference*. HCI International, Crete, 559–570.

Djaouti, D. (2019). *La préhistoire des jeux vidéo*. Ludoscience, Talence.

Kajastila, R., Holsti, L., Hämäläinen, P. (2016). The augmented climbing wall: high-exertion proximity interaction on a wall-sized interactive surface. In *Proceedings of the 2016 CHI conference on human factors in computing systems*. ACM, San Jose, 758–769.

Wolf, M.J.P. (2008). *The videogame explosion: A History from PONG to Playstation and Beyond*. Greenwood Press, Westport.

8

Projection Mapping and Photogrammetry: Interest, Contribution, Current Limitations and Future Perspectives

8.1. Introduction

Photogrammetry is a process that allows us to reconstruct an object or a place in 3D from photos. It has undergone a recent and important development that makes it possible to consider applications in the field of projection mapping, in particular for the creation of animations, the positioning of projectors and the preventive implementation. While these advances are currently of application interest mainly for object projection mapping, there is no doubt that in the short term they can be deployed for increasingly important projection mapping project. After a brief overview, this chapter details the interest that photogrammetry can have for projection mapping and makes its contribution to it: a shooting device optimized for photogrammetric reconstruction, easily usable for object projection mapping. The second part considers the short-term developments in photogrammetry and the implications they will have at different levels of projection mapping design and production.

8.2. State of the art

The principle of photogrammetry has been known for a long time. From the parallax differences between different photos of the same subject, it is

---

Chapter written by Nicolas LISSARRAGUE.

possible to obtain a topographic representation. While Alberti's famous *contructione leggitima* allows us to convert a plan into a perspectivist representation, photogrammetry does the opposite, but with an additional difficulty: it cannot know the scale of objects from a single photo. Since a large distant sphere may appear the same size as a small nearby sphere in a perspective view, photogrammetry requires several photos taken from different angles to cross-reference information and determine the objective size of the objects represented (also called metrophotography). Colonel Aimé Laussedat was the first to apply this principle to the facade of the Hôtel des Invalides in 1849. Until the advent of digital technology, photogrammetric reconstruction is carried out with pairs of photos – often obtained through stereoscopy – using a complex restitution device. Widely used in architecture, archaeology or topography, the process is long and its accuracy limited.

The evolution of information technology makes it possible, in the first instance, to virtualize the restitution device. Software such as Canoma or ImageModeler[1] allow you to take precise measurements of the same area on several photos. But the use of algorithms based on the mathematical principles of triangulation not only provides a two-dimensional topographic representation, but also creates a 3D representation of the subject being photographed. Three areas of digital technology, as they progress, influence digital photogrammetry:

– increasing the resolution of photos;

– increasing the processor power;

– increasing the efficiency of computer vision[2].

This last field facilitates an important evolution, which is intensively exploited by current software: the automation of the identification of the same area in different shots, the last operation that still had to be carried out manually, results in the complete automation of the photogrammetric reconstruction process. The ever-increasing definition of digital cameras leads to greater accuracy of the 3D model. Finally, photogrammetry, both because it uses triangulation calculations intensively and because it must

---

1 Softwarepublished respectively by MetaCreation and Realviz in 1999 in their first version (www.canoma.com/).
2 Computer Vision: a branch of artificial intelligence that seeks to develop a computer's ability to analyze and process images automatically.

exploit large volumes of data, benefits directly from the increase in processor power. Automation, process acceleration and accuracy of the result are now leading to new uses of photogram measurement, especially for projection mapping.

8.3. Photogrammetry for projection mapping

Recently, new players in the digital industry have developed workflows involving photogrammetry. The Game Developers Conference 2018, for example, gave precedence to the use of photogrammetry in games[3]. Released in 2017, RealityCapture software has been optimized to take advantage of the computing power of GPUs (their ability to do parallel processing makes them particularly effective for calculations involved in triangulation and computer vision). Not only has the reconstruction of the 3D model been significantly accelerated compared to previous generations of software, but it also makes it possible to manage very large quantities of images: the 3D reconstruction of the city of Banská Štiavnica or St. Elizabeth's Cathedral in Kosice was made from more than 100,000 photos[4]. To create a virtual double of an object or character, companies have developed installations that optimize the shooting process. Where photogrammetry was originally satisfied with a couple of photos to reconstruct a topographical plan, it is now possible to use devices that mobilize dozens of digital cameras. For example, Ten24 has a platform with 170 synchronized cameras![5] Gnomon has developed a similar device... in a truck[6]. In both cases, a few seconds are

---

3 See for example: "Automated Photogrammetry to Game Res Pipeline" (www.youtube.com/watch?v=XE7MRHEhPGc), "Creating realistic environments for Forza Motorsport 7: Photogrammetry and laser scanning" (www.youtube.com/watch?v=ZGQTH-IuEFc), "Photogrammetry and Star Wars Battlefront" (www.youtube.com/watch?v=U_WaqCBp9zo) or "A VFX Workflow for Real-Time Production with Photogrammetry, Alembic and Unity" (www.youtube.com/watch?v=xKWQBSnhExM). Ubisoft also reported on an interesting test: it asked two different teams to model a 3D object in high definition. The first team produced the *ex nihilo* model on a 3D software in 14 days. The second team, to obtain an equivalent result, took only four days using photogrammetry.
4 The emergence of many projects to rebuild cities through photogrammetry illustrates the new possibilities offered by the increase in computer power and software optimization. See: www.youtube.com/watch?v=b6E7F3Slbys or www.youtube.com/watch?v=K7onvPSwhbQ or https://www.acute3d.com/city-mapping/.
5 See: http://ten24.info/3d-scanning/ and the ScanLab company's system (https://scanlab.ca/) using 132 cameras.
6 See: https://vimeo.com/265247923 and www.facebook.com/thescantruck/. Even more recently, Intel presented the first attempt at video photogrammetry (or volumetric cinema) at

enough to photograph a subject from all angles and with optimized lighting, simplifying and accelerating the reconstruction process. The value of these developments for projection mapping is considerable. Photogrammetry allows us to reconstruct a virtual double of a subject, offering a simple and efficient way to preview a projection on a volume. In the case of object mapping in particular, being able to manipulate the object in 3D makes it possible to best prepare the placement of the projector, to visualize the result of the projection in a virtual way to better optimize it, but also to be able to extract data from the 3D model that can be used to perform the projection itself. But for the quality of the reconstruction to be satisfactory, it is necessary to have many photos of the subject, taken from all angles. The devices mentioned above require a heavy and imposing investment, which makes them difficult to use for object projection mapping – which is why they are mainly used by the major players in the gaming and film special effects industry. However, it is possible to adapt the principle of their operation to a simpler and less expensive device.

8.4. Contribution: an automated imaging device for object photogrammetry

If photogrammetric reproduction sets require so many cameras, it is essentially in order to be able to reproduce people: to be accurate, the reconstruction must be based on photos of a perfectly motionless subject, which is practically impossible if the shooting lasts several seconds. But if we limit ourselves to the reproduction of objects, it is possible to simplify the device, the need to take all the shots at the same time no longer being decisive. It is on the basis of this observation that the project of a system coordinating a rotary table and several cameras was developed, based on a project initially conceived by Vladimir Olegovich[7]. The main characteristics sought were:

CES 2018: by filming a scene with 96 high-resolution cameras, it was possible to reconstruct it over its entire duration in 3D. The result is available at this address: www.youtube.com/watch?v=9qd276AJg-o. For further explanations on the process, see for example: www.usine-digitale.fr/article/intel-a-mis-la-video-volumetrique-a-l-honneur-au-ces-2019-avec-un-hommage-a-grease-et-une-demo-de-mimesys.N792874.

7 See: http://makerdrive.org/ (in Russian). This project was initially developed around an Arduino to create 360° photographic windows for the Internet. The more recent versions of the project are based on an Espruino, which is not open source and is significantly more expensive.

– a versatile system, easy to transport, capable of being declined in different sizes in order to be adjustable to the weight and dimensions of the objects to be captured;

– an open source system that is accessible and as inexpensive as possible;

– a system that is simple to build and can be realized with minimal knowledge of electronics and programming;

– a reliable, simple and quick to use system;

– rotary table control options optimized for photogrammetry.

The first version of the rotary table meets most of these specifications. The rotary table itself, with a diameter of 34 cm, combined with a NEMA 17 stepper motor can handle a load of more than 50 kg. Its size can be easily modified (versions of 25, 34, 50 and 80 cm have been produced) and all the parts that make it up can be machined with a laser cut. The control box allows you to set the speed, acceleration, number of stops per revolution and pause time per stop. It is based on an Arduino, easily accessible components that do not require soldering, and allows up to three cameras to be triggered simultaneously. In order to optimize the quality of the shot, the rest of the device includes, on the one hand, a matt black background (to prevent photogrammetry software from attempting to reconstruct the set around the photographed object) and, on the other hand, a lighting system using an annular flash and a dual polarization of light[8].

Figure 8.1. *The rotary table (source: Nicolas Lissarrague). For a color version of the figures in this book see, www.iste.co.uk/schmitt/image.zip*

---

8 This type of lighting is inspired by photography in dentistry: it considerably reduces the reflections, which distort photogrammetric reconstruction. It is also used to obtain shots without shadows and, later in the reconstruction process, a texture that is easier to use.

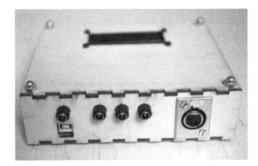

Figure 8.2. *The control box (source: Nicolas Lissarrague)*

After several tests on objects of different sizes and materials (plaster bust and statue, polystyrene block, shoes, log, metal pot, etc.), it is possible to make a first observation. The system is fully satisfactory in terms of the quality and speed of the reconstruction[9] (Figure 8.3).

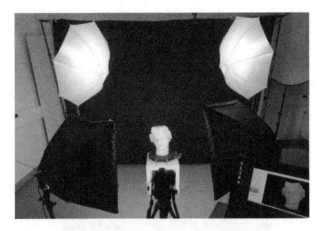

Figure 8.3. *The complete system during a shooting session (source: Nicolas Lissarrague)*

In the first configuration, using three cameras, it takes three minutes to get 54 pictures of the object: the turntable rotates 20° between each picture, pauses for 15 seconds (to give the flash time to recharge) and then moves to the next position, repeating the operation 18 times. The cameras are positioned to simultaneously photograph the object from the front, from

---

9 All sources and explanations necessary to reproduce the device are available at this address: https://moodle.uphf.fr/course/view.php?id=1166.

Projection Mapping and Photogrammetry   133

below and from above. It takes about three hours for the entire process, from the setup of the devices (rotary table and cameras) to the final reconstruction of the object in 3D.

In the second configuration, using five cameras, it takes 15 minutes to get 180 pictures of the object. It takes about five hours to get the object rebuilt – the process time is essentially extended by the longer rebuilding time because it is based on a larger number of photos than in the first configuration.

Figure 8.4. *Contact sheet of some of the photos taken during a shooting session (source: Nicolas Lissarrague)*

If the results obtained are satisfactory, experience shows that improvements in the ergonomics of the device are desirable: management by remote control is impractical, the cable connecting the control box to the rotary table is often annoying, feedback is often missing (on remaining time, reminder of parameters, etc.) and several options could be added (including the possibility of continuously rotating the rotary table and triggering video cameras rather than cameras). A version 2.0 of the device is currently under development, which will integrate all these modifications, will enable up to nine cameras to be triggered simultaneously and will be controllable via an Android application[10].

Figure 8.5. *3D model obtained by photogrammetric reconstruction (source: Nicolas Lissarrague)*

Once the 3D model is obtained from the photogrammetric reconstruction, previewing a projection mapping work becomes very simple. First of all, it is necessary to simplify the mesh size, which is often very dense and too heavy for a real-time display, either with the decimation functions existing in photogrammetry software[11], or with a specifically dedicated software[12], or with the functions integrated in the main 3D software[13]. The simplified 3D model must then be imported into 3D software: by rendering the object with a camera, the image of the 3D model is obtained at a precise angle, which

---

10 Version 2.0 is currently being developed by students at ISEN (Université catholique de Lille) under the direction of Laura Saini – for whom I am grateful. The sources will be accessible on the link mentioned above.

11 This is the fastest solution when you want to manipulate the 3D model and its texture. The other possibilities involve two additional steps: creating mapping coordinates for the simplified model, and then transferring the texture of the complex model to the simplified model (or *baking*). On the other hand, if it is a question of using only the 3D model without texture, the three possibilities are equivalent.

12 As InstantMesh (free) or TopoGun.

13 3DsMax, Maya, Cinema4D, Blender, Houdini, ZBrush all have this type of functionality.

can then be projected back onto the model via the same camera. The image can simply be used to delimit its silhouette in order to use it as a mask for projection, or it can be used to create the projected animation. It is at this level that the 3D preview obtained by photogrammetry has three major advantages:

– optimization of projector placement: the 3D projection makes it possible to clearly show the areas that are not covered by the projection;

– 3D data recovery with usable rendering for animation: most 3D rendering engines (Arnold, MentalRay, VRay, RenderMan, etc.) can extract data such as the orientation of the model's faces relative to the projector (normal map), relief areas (*convexity* and *cavity map*), distance from the projector (ZDepth map), the shape of the 3D object (Alpha map) or simply its original texture (Diffuse map). These textures can then be used to create the animation to be projected;

– visibility of overlay areas when multiple projectors are used: by projecting test patterns from multiple locations, overlay areas can be anticipated and optimized.

Figure 8.6. *Diffuse map (source: Nicolas Lissarrague)*

Tests on different objects show that it is possible to prepare the positioning of a projector and perform a simple animation in a few hours – five hours for objects with a simple spatial structure and 8 to 12 hours for more complex objects or multiple projections. The use of a turntable and automated shooting significantly speeds up and simplifies the creation of an animation and the placement of one or more projectors for projection mapping.

Figure 8.7. *Alpha map (source: Nicolas Lissarrague)*

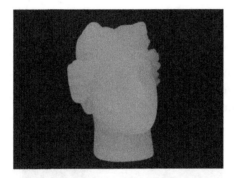

Figure 8.8. *ZDepth map (source: Nicolas Lissarrague)*

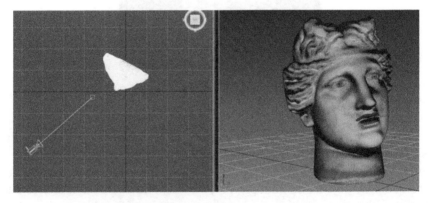

Figure 8.9. *Projection of the Diffuse texture retouched (left: view from above of the position of the 3D object and the projector; right: visualization of the result) (source: Nicolas Lissarrague)*

8.5. Current limitations and future prospects

While the tests carried out are encouraging, in that they imply both simplification and speed of the process, they are not yet integrated into a smooth and coherent workflow. Between the shooting and the projection itself, many software packages are required and the transmission of data from one to the other is still limited. However, while the main projection mapping software only operated in a two-dimensional mode, most of them now include features to import and manage 3D models to preview the projection.

The joint presentation by Nvidia's RealityCapture's and Unity's development team at the Game Developers Conference 2017 shows the desire for an ever-increasing integration between the stages of 3D shooting and reconstruction[14]. If we add the recent possibility offered by Millumin software to easily import 3D scenes from Unity[15], we can guess that future evolutions will tend towards a much more unified environment than at present.

On the other hand, the tests were carried out on small volumes, proving the interest of the device for object projection mapping, but not on large surfaces for more monumental projections. While the use of photogrammetry is technically possible in this case, it poses almost two problems: the first is the difficulty of access to the high parts of an architecture for photography, which frequently requires the use of a drone, but the development and ease of use of current drones tends to reduce this problem more and more. The second, more problematic, is the much larger amount of data to be processed: several hundred, if not several thousand, photos are needed for a large-scale reconstruction, which multiplies the shooting and calculation times[16]. While the recent possibility of using videos instead of photos partly removes this obstacle, the computer power required remains a limitation.

---

14 A transcript of the conference is available at this address – the video is only available with a paid subscription: https://developer.nvidia.com/gdc17.
15 Other software such as Designer or Previz have also developed possibilities for data exchange with 3D software.
16 This is called "videogrammetry". Some software such as Zenith 3D is able to analyze the video sequences and extract the most relevant images before performing the reconstruction.

Photogrammetry still suffers from limitations that do not allow it to be used for all types of projection mapping. Of course, recent developments in this technique have made significant progress, but they are still insufficient to allow a smooth and integrable use in all the different types of projection mapping. However, developments in other areas of digital technology will quickly contribute to driving new developments. Three of them deserve to be monitored.

The first is the possibility of using decentralized calculation for reconstruction. Already functional for applications such as speech recognition[17], its adaptation would speed up the reconstruction process, even when large volumes of data are required.

The second concerns the evolution of digital vision algorithms. Their optimization allows us to consider the possibility of a photogrammetric reconstruction carried out in near real-time, of which ARCore[18] gives an overview.

The development of mixed and augmented reality tends to converge with that of photogrammetry in that it is necessary to model the environment in 3D and then superimpose it in real-time on the real environment in order to integrate virtual elements – this monitoring, or *3D tracking*, is based on triangulation calculations, etc., just like photogrammetry. It is also possible to imagine that in the long run, mixed reality will develop individualized projection mappings.

The third is the growing influence of the contributions of artificial intelligence and machine learning for image analysis and pattern recognition. Not only does their contribution to computer vision improve accuracy and speed, but the automated detection of complex patterns (silhouette, face, window, doors, etc.) would make it possible, for example, to achieve

---

17 The SIRI speech recognition application on iOS systems is the best example, which transmits speech recording to remote servers for analysis. The booming *cloud computing* market is likely to contribute to the development of remote and real-time photogrammetry computation.

18 See https://developers.google.com/ar/ but also https://developer.apple.com/arkit/.

complex masking effects for projections,[19] or to be able to perform projection mappings on moving parts[20].

In the end, photogrammetry may only be the currently visible part of the growing influence that digital developments have on the design, creation and production of projection mapping. But the examples mentioned all show the extent to which this will be the challenge of major changes in the near future.

8.6. References

Egels, Y., Kasser, M. (2001). *Digital Photogrammetry*. CRC Press, Boca Raton.

Kraus, K., Waldhäusl, P. (1997). *Manuel de photogrammétrie: Principes et procédés fondamentaux*. Hermès, Paris.

Luhmann, T., Robson, S., Kyle, S., Boehm, J. (2013). *Close-Range Photogrammetry and 3D Imaging: 3D Imaging Techniques*, 2nd edition. De Gruyter, Berlin.

Wolf, P.R. (2014). *Elements of Photogrammetry with Application in GIS*. McGraw-Hill Education, New York.

Zemichael, Y. (2012). *Digitalization of objects by photogrammetry*. Lambert Academic Publishing, Sarrebruck.

---

19 The presentation at the 2018 ECCV of the results obtained with the DensePose-COCO, for example, shows that the ability of a computer to recognize one or more human figures in a video has improved considerably: http://densepose.org/. Other systems, such as CycleGAN, also demonstrate the progress made by *machine learning in* image analysis and processing (https://junyanz.github.io/CycleGAN/).

20 See www.youtube.com/watch?v=-bh1MHuA5jU. It is easy to imagine what a projection device of this type coupled with the image analysis capabilities mentioned in the previous note would allow: a projection mapping to be made on any type of mobile subject (I thank Ludovic Burczykowski for having helped me discover this source).

9

Points of View: Sound, Projection and Interaction

9.1. Sound creation projection mapping, a real composition of sound

9.1.1. *Introduction*

Although sound and video are two distinct artistic disciplines, it is not uncommon for the sound part to be relegated to the background in projection mapping projects. It is sometimes just a simple musical accompaniment with an explanatory voiceover. The soundtrack must be an integral part of the work and bring its own narrative by putting distance and rupture in front of the visuals. Like video, which results from the assembly of a multitude of images, layers and masks, it is necessary to think of it as an agglomeration of sound layers constituting a sound landscape.

Sound is always a phenomenon following an event. It is the trace and the residue. It is also the resonance of an act, a movement, the encounter between two materials. Its energy impacts and affects the meaning of what we see because it gives a complementary reading. Indeed, sound enriches this off-field space impossible to show through the image. It develops the contours of the world that the work seeks to create by letting the spectators appeal to their imaginations. Indeed, everyone interprets what they hear according to their own experience and culture. The spectator relativizes what he sees and the visual impact of the projection is accentuated.

Chapter written by Jérémy OURY, Ludovic BURCZYKOWSKI and Marine THÉBAULT.

Projection mapping is the projection of visual content on a surface other than a cinema screen in order to create a virtual space based on the geometry of the medium. This practice is essentially oriented towards the image because it is above all a projection of light on a surface. Thus, the choice of this one sets the basis of the project, and a majority of projection mappings are called architectural because they are made on buildings. It is quite common to see projection mappings where the sound part is only a lively and rhythmic music, sometimes associated with a voice-over accompanying the video. The heritage of fireworks and the spectacular events of "sounds and lights" partly explains this emphasis on the visual over the sound. Unfortunately, some projects are then reduced to the accumulation of breathtaking 3D effects without any narrative or artistic aspect.

However, other initiatives offer the opportunity for sound to be the creative and narrative driver in combination with visual content. This can be particularly true when the projection medium has a strong sound identity such as churches, ports and industrial buildings or, on the contrary, neutral like a white wall or an abstract form. In the first case, the reuse of sound effects from the place creates a sound universe connected to its history, while in the second case, sound can become the basis for reflection in the project. The two media, sound and visual, each carry a part of the narrative, making the work coherent and strong.

Can sound claim a place other than a complement to a projection mapping work? Can sound be the first field of investigation in the creation of a visual work?

This chapter will try to answer these questions and show, through a reflection on its use in other artistic forms, the importance of involving the sound part in the process of creating a projection mapping work. This reflection will be completed by an analysis of the works of Telenoika, AntiVJ and OnionLab and based on interviews with Thomas Vaquié on two of these works within the AntiVJ collective, and Santi Vilanova from Telenoika.

9.1.2. *The place of sound*

9.1.2.1. *The contribution of other arts*

Several artistic forms have a strong correlation with sound such as theatre, radio and music.

In the theatre, there was a behind-the-scenes technician who created off-screen spaces and sound occurrences with, for example, a metal plate to play the sound of thunder. The arrival of analogical and then digital technologies with recording, reproduction and amplification systems has developed the place of sound, which has become a real companion of the actors and a real complement to scenography. The sound engineer acts with the set designer from the beginning of the show's creation by working on the spaces and off-field. Sound in the theatre is not thought of as a frontal projection, but as multiple sound sources arranged in the play space, behind the scenes or around the audience to have an immersive sound architecture.

Although listening is personal and unique depending on the spectator's placement in space, it has a strong influence on the imagination. By playing on the texture of the sound, its grain, its height, its blur, its density, its homogeneity or its plasticity, it is possible to offer the audience another interpretation of what happens on stage. Soundtracks are constructed in layers and by combining different sounds. They are played independently in real-time to keep pace of the actors' rhythms and the events taking place on stage. The writing is not fixed in time, which implies that a sound engineer interprets it at each performance.

Like theater, radio is an interesting medium that appeals only to our ears. Sound is therefore at the center of the creative process and is the main focus of the work's writing. Accompanying one or more voices, the soundtrack, close to that of the theater, combines realistic or unrealistic sound effects with occurrences, abstract layers and music. Played live, it accompanies the voices of the readers. However, if it is carried out during a software editing, it is then frozen in the temporality of the work.

Another interesting contribution is that of music. It is an abstraction that plays with energies and rhythms by imposing a linear duration. Contemporary music has developed the use of sound effects as instruments. The sound of these "noises" has the power to remind us of the elements of everyday life. Some artists like Stockhausen also imagine spatial experiments by placing several sound sources in space to place the listener in total immersion. The writing of the soundtrack therefore also involves the choice and placement of the public address system and the spatialization of these sounds.

9.1.2.2. *Synchronism and synesthesia*

Cinema also has a special relationship with sound. With the emergence of techniques related to sound recording and reproduction, silent cinema has been transformed into sound cinema. The term synchronism comes from the action of coordinating, in post-production, the image with the sound associated with it. Indeed, the two media are not always recorded on the same medium or at the same time. By extension, synchronism also means associating a sound with a movement in order to give a perception of a link between the two events. In modern cinema, the whole sound part, from voices to sound effects, is completely redone in post-production. The effects of real life are sometimes saturated by the addition of artificial sound effects to any moving element. Thus, the synchronism to which the cinema clings would result in the following, whereby the sound confirms the image and the image justifies the sound. This often implies that the sound doubles what the video already says.

Several filmmakers have proposed a different use of sound by seeking "non-coincidence with visual images" to create what they call "vision images" and "sound images" where sound acts as a counterpoint to video (Eisenstein *et al.* 1928).

If synchronism is the association of a sound with an image, synesthesia comes from the idea of merging perceptions. It is a neurological, involuntary and permanent phenomenon in which a person combines several senses. For example, seeing sound and hearing light or associating a number with a color. This remains a notion of subjective perception. The development of abstract art has built a new perceptive and sensitive experience for the spectator through artificial synesthesia. Indeed, digital art has notably introduced the fusion of different media in the same language to obtain an audio-visual combination resulting from the same writing. More than synchronism, the work is generated through a common sound and visual language. In this generative process, sound either becomes the driving force behind visual creation or is directly associated with it.

9.1.2.3. *Areas for reflection*

Art is above all an experiment. By bringing the sound to the same level of writing as the visual, another meaning can emerge from their combination in addition to what they carry themselves. If the sound suggests information that is different from what the video shows, there is an additional axis for reading the work.

It is possible, for example, to imagine sound sources off-screen to enrich the space. The sound thus widens the narrowed image around the projection medium. Being in a sound economy can be a possibility because the desire to hear is important. Indeed, by working on the loss of sounds, removing elements from them and streamlining them, we give more space to the spectator's imagination since a few sound elements are enough to appeal to their memories.

The desynchronization of two dissociated elements placed side by side creates confusion and questioning. Thus, it can be interesting to work in a free association of independently created visual and sound elements. The artistic experience of associating a sound creation by the composer Cage with a choreography by the artist Cunningham produces effects of reference, correspondence and chance, whereas the two elements exist independently. Our brain will unconsciously search for synchronization points between different media.

Parallel writing of sound and video is also an option. On some sequences, the video adapts to the sound with visual elements designed using the sound rhythm. On others, especially the slower moments, the video takes precedence over a sound atmosphere.

The use of music often induces a distorted narrative because of its repetitive nature. It imposes a linear duration on visual effects and their repetitions follow the construction of melodic sentences. To avoid these automatisms, the soundtrack must be specifically designed for the project and combine sound effects and sound ambiences to avoid a lack of connection to the visual.

Finally, it is not uncommon to find a voice-over on "historical" mappings. This voice imposes an overloaded truth and often comments on what the image already offers. Nevertheless, on a poetic text, like the latest *Loneliness* creation by the Russian studio Sila Sveta, its use is relevant. The voice then takes on a narrative aspect and gives a different reading of the projected images. Video is derived from the rhythm and melody of the voice and not from the meaning of the words.

9.1.3. *Analysis of works of art*

The approach of the AntiVJ label is to produce audio-visual objects with positions on the choice of images and sounds. Reflections on the aesthetics of the project lead to a narrative or a poetic anchoring of the work. Thomas Vaquié, the sound designer, explains the creative process of the *Omnicron* and Orgue de Saint Gervais works.

9.1.3.1. *Creative process on the work Omnicron, AntiVJ, 2012*

The Omnicron project was born from a commission for an *in situ* creation. Indeed, Karol Rakowski of the Polish collective Spectribe wanted the AntiVJ collective to make a permanent audio-visual projection on the Hala Stulecia dome in Wroclaw. This imposing building, built by architect Marx Berg in the early 1900s, is entirely made of concrete. The material and physical constraints were obvious from the first visit to the site, namely the possible projection surfaces, the placement and number of projectors, the resonance and reverberation of the space, and the placement of sound speakers. These elements have a direct implication on creation.

The writing process begins with a definition of the themes and narrative structure. In view of the history of the building, the project also addresses gigantism and time. We find the fantasies and arrogance of the architect imagining his imposing building in a future time. The projection surface being complex and limited, figurative contents and virtual light effects are impossible. The visual aspect will be based essentially on simple and minimalist elements, consisting of lines and flat areas of light.

Thomas Vaquié explains this approach: "On the one hand, let these white forms, this light energy, be heard in order to create an audio-visual object, and not an image accompanied by music or the opposite. The idea is to create the illusion that sound is produced by the projected lights. On the other hand, try to bring a second layer of reading, perhaps a little less frontal, a little more poetic. This aspect requires a harmonization of sounds and a rhythmic development of these sounds." On *Omnicron*, his creation is a kind of echo of the mixture of pride, anger and satisfaction of this man who leaves the trace of his existence within the walls of this imposing building.

The music borrows colors from the post-Romantic music of the late 19th Century, with the use of power and, sometimes, the dissonance of brass and large sections of orchestra mixed with a layer of electronic sounds, symbols of the emerging industrial era. Thomas Vaquié also uses *theremin*: the sounds of this electronic instrument invented in 1919 represent the idea of the future at the beginning of the 20th Century.

Concerning the work process, Romain Tardy and Guillaume Cottet created a first introductory part where Thomas set sound effects in synchronization. This part is less rhythmic than the next, where Romain used the sound composition to create the visual animations. In both cases, there were several exchanges between the two artists to specify synchronization points and reinforce certain sound and visual movements. Working in parallel improves overall coherence and the progress of the creative process.

What marks this project is the introduction of a musical phrase systematically challenged by visual ruptures that cause an arrhythmic revival of sounds. Thus, the musical loops give rhythm to the work without becoming the main element.

9.1.3.2. *Analysis of the work Orgue Saint Gervais, AntiVJ, 2010*

The analysis of this creation is interesting because the projection is made on a sound object: the organ. Sound is therefore at the center of the creative process. The project originates from Yannick Jacquet's desire to experiment, after a first attempt at VJing with real-time image projection during an organ concert, with an audio-visual project on this instrument in a church. The temple Saint-Gervais is made available to this creation at the invitation of the Mapping Festival in Geneva.

The project begins with the discovery of the place and the meeting with the organist. The resonance of the church is the main constraint and the discussion leads to the choice to play the organ live with a soundtrack, then, in the second part, a piece by Arvo Pärt. The objective is the integration of the organ with electronic music while respecting the minimalist Protestant aspect, which is reflected here by a holy place with a white wall, without decoration. Thomas Vaquié decided to use synthesizers to reproduce the organ system which, from a keyboard and zippers, produces a panel of very different sounds such as those of percussion or others close to a human voice.

The creation is initiated by the musical composition then by an audio reactivity work to generate a part of the visual. It was then necessary to imagine video as an instrument since it was the sound that controlled it. The rendering on site was chaotic because the resonance of the place gave interference in the soundtrack. Thomas had to work on the spot on his part by simplifying the soundtrack as much as possible, especially at the bottom and in the high notes. He kept the synchronisms while spacing the sounds as much as possible to give them time to disappear into space. This shows that an audio-visual mapping creation must be conceived for a place in an ephemeral way and requires an *in situ* creation or adaptation work, especially in complex acoustic and architectural spaces.

9.1.3.3. *Analysis of Telenoika's Kernel work and Onionlab's Evolution*

Two Spanish works, the mapping of the Telenoika collective projected during the Kernel festival in Italy and *Evolución* by Onionlab, show the importance of sound for the project concept.

The work of the Telenoika collective impresses by the very close link and perfect synesthesia between sound and visual elements. It is a piece that involves a great dynamic between silences and massive percussive sounds associated with a great rhythmic amplitude, starting with a very slow evolution of the fractal visuals at the opening until the final stroboscopic sequence.

Santi Vilanova, the sound designer, details the process of creating the soundtrack and the visual elements in parallel. This was done collectively with a team of musicians and a team of video creators. Everyone produces their ideas and then the sounds are transformed into images and the images into sounds. Telenoika often uses this function to guarantee a perfect unity between sound and video as if they came from a single source and the same writing.

Playmodes Studio, Santi's second collective, also explores the links between light and sound through immersive LED installations such as *Porta Este-la* and *Beyond*. In these installations, the studio adds a spatial dimension to sound creation by imagining its use in space by having more broadcasting points than a frontal projection mapping.

On the creation *Evolución* by OnionLab, created during the Mapping Festival in Geneva and then adapted to the Signal Festival in Prague on the Santa Ludmila church, the sound part counterpoints the visuals. It then takes the form of a drone slowly evolving to bring us to a sensation of trance. This sound continuum, which only slightly emphasizes video crackling, takes us from the beginning to the end of the work to hypnotic visuals that do not have the same rhythm as the soundtrack. The few rare points of synchronism desired by the creators or interpreted as such by the spectators make it a singular work where video and sound are in opposition to each other to express in an abstract way this feeling of time that passes and the alteration of the building.

9.1.4. *Conclusion*

Sound, like many other artistic practices, often has to struggle to find its place. It is often thought of as the last step in the creative process when it may offer the most imagination, illusion and narrative.

Like sound creation in the theater, it is not a musical accompaniment, but a multitude of layers reproducing a real imaginary space through several sources. Projection mapping should therefore multiply the writing connections between the two media, sound and visual, that compose it. The field of possibilities is wide because sound can be designed in parallel with video, be the initiator of the project or be the result of a synesthetic generation with the visual. The processes of visual and sound creation can therefore be considered in parallel or by determining sound as the structuring element of the project.

In the context of outdoor projection, although urban noise imposes constraints on sound levels and the placement of sound sources, there are still many artistic possibilities for creation provided that the sound is considered as equal to the visual. The sound creation of projection mapping offers more opportunities for indoor projections, away from our daily noise, in terms of dynamics and subtlety.

But in order to give sound its place in this art of illusionist projection and to bring it to the same level of importance as video, the starting point would be to systematically associate a sound creator with the artistic project and, perhaps, to rename the practice of projection mapping as video-audio mapping or audio-visual mapping.

9.2. Projectionist: a profession according to Pascal Leroy

9.2.1. *History*

Pascal Leroy has been using video projectors for more than 20 years. His official title is projection technician or projectionist, although over the time that he has been practicing this profession, he has become so visible to his interlocutors that he deals with projection that they simply call him Pascal. He has been working on various things and doing about a hundred projects a year for the past 20 years. Pascal Leroy does everything that can be done with projectors, he says, which probably increases the quality of his expertise. He can use what he learns in one context to put it in another. "I did some unsatisfactory things too," he says. In addition to projection mapping, these are films, film concerts, seminars, events, conventions, etc. Pascal Leroy is an expert in the sense that he brings something by a solid experience, broad and specialized: whether in consulting from the design phase or in technique during the diffusion. But it would be simplistic to suggest that he is only a technical person: he has a sharp eye and artistic sensitivity, he says, and overflowing imagination is part of his job. He is above all from the house of the performing arts.

Pascal Leroy was passionate very early on but without wanting to become an expert at first. Already as a child, he made "shenanigans" with 8 mm projectors. He had discovered that you could project a camcorder onto a ceiling in the dark by acting on his eyepiece and using it as a projector. All this was written somewhere, perhaps, he wondered, as a foregone conclusion. Then, Pascal Leroy used oil-immersed projectors that no longer exist today. We didn't talk about mapping in the beginning and since he was projecting before projection mapping appeared, he said he saw it being born, metamorphosing, mutating. Twenty years ago, the tools available were not suitable for mapping, he says. This is not new and Jean-Michel Jarre was already doing it, but the spotlights were not powerful enough. The first mapping he participated in was with the *1024 Architecture* collective. Together they wiped a lot of plaster in his opinion: "We didn't know if it would work..." But it was revolutionary. *1024 Architecture* used white threads because of their origin as architects, he thinks: they drew dimensions on design plans at school. In doing so, they had chosen to explore a small part of what is possible but it fulfilled its role very well. Then, Pascal Leroy discovered other forms. He admits that at first he was convinced that projection mapping would be an ephemeral fashion. But today, he notes that projection mapping has spread: large, infinitely expandable, in motion, in various contexts, historical, spectacular, etc. In fact, he suggests, it's like fireworks: the whole world loves it and we do it again. A little like champagne too, which sparkles and is good, plus you never get tired of it, even if you know the taste.

9.2.2. Identity and tastes

Pascal Leroy says he loves color and is carried away by the sound and visual combination, when "it is a question of fusion and that the trend develops. It is also touching to see many people," he says. He sees two types of profiles in designers: those who stay in place and those who go further. One day, the projection on the facade will no longer be enough, he thinks, it must be mixed with fireworks, smoke for example, physical persons, interaction, lights, etc. Connect several buildings, add dimensions, use more ingredients, such as extra objects. You have to dare, be a little crazy, find new uses. According to Pascal Leroy, we see places where projection mapping is running out of steam: it is complicated to innovate, and difficult to surprise people. Yet surprise is the keystone, the guiding principle, it is where there is a reaction and where chemistry takes place. Being surprised remains the common denominator, that's what creates emotion. But be careful not to burn all your cartridges, there may be overbidding, inflation. You can keep your secrets, not tell everything, to keep the surprise or keep ahead, and not do what others have done.

9.2.3. Art and technology

Pascal Leroy argues that the projector is the first link in the chain of a project with projection, and hopes that technology will not be a hindrance: "tell me, and I will do it." Unfortunately, there is no school that trains projectionists outside the CAP as projectionist operators. They only make films, which is a very focused, strict and narrow way of using the projector. It is merely a question of following procedures and there is no additional information to be provided by the person making the projection.

"There is always a little superstition in the show," he also notes, and even a lot. But there are more of them among artists than among technicians. In technique, a maniacal side is more represented, which is not quite superstition. It is then a question of loving symmetry, order, and rationalizing to achieve control and not leaving room for chance by quickly identifying things. As soon as it gets complicated, we have to codify and rationalize according to Pascal Leroy.

Technique is the painter's palette and Pascal Leroy likes to propose other colors whose artists could not even imagine their existence. He tries to master the tools, then to have an artistic look at the technique. "It is not well understood that a technician can work in an artistic way," he says, "as if they were denied subjectivity." However, the technician offers tools to the artists and the artists also have their own methods, so they are linked. It is the combination of talents that brings a quality: technical ideas give artistic ideas, tools take artists further and further.

9.2.4. *Limitations*

However, the necessary darkness is one of the notable problems. "Mapping in the middle of the day would be nice," he imagines, "but we lack power. And you can't project beyond the medium," he continues: "Projecting the image projected into a building from a building would probably be the same. But we have just planted the first tree in a forest, the playground is large and diversified since the frame was removed from the screen. Can we map the moon? We can find a limit to mapping the planets. God maps with light, we only reproduce that," says Pascal Leroy. Nature, vegetation is mapping on a planet; the moon changing color is mapping... Pascal Leroy says he will stop when he has mapped a planet. Then he'll be done. He wants to travel far away and there is still so much to do that he hopes to be part of.

9.2.5. *Projection mapping and cinema*

Cinema is the Grail of the image for Pascal Leroy. He is untouchable and infinitely respectable. But cinema is formal and framed, in all senses. Rigid. At the cinema, we bring in different cooks, but they always cook in the same frying pan. The mapping removes the frame: we decide which support is part of the work. In mapping, it would be a narrow-mindedness to consider the medium as a "nothing". Besides, *mapping* may not be the best word, he thinks. We could talk about *skinning*, another skin, or dressing, even if it is non-tangible or virtual. Or even to undress. And the skin is not on a void. In projection mapping, we say "we're going to change the rules", at first, we say "wow, s***", and then it encroaches on a lot of fields. There is the movement of the medium that is possible, the volume, a surface that is not necessarily opaque, not necessarily hard... The playing field becomes gigantic, while it is more difficult to add a dimension to the cinema. We try 3D cinema with stereoscopy, or fans to make 4D, outdoor projection, but it remains a rectangle. Cinema is only the reading of the work. Mapping is about performance. There is the weather, the danger, a fragility that comes into play. Pascal Leroy makes it a maxim that projection mapping breaks chains and frees the image. Projection mapping is not only about images; it is like an outcome where genres are combined and mixed.

9.3. Interactive projection mapping by Anne-Laure George-Molland

9.3.1. *Enter interactivity to make it exist*

There is interactivity as soon as a work evolves, is transformed in contact with the public and its environment. Artists can use the ability of computer systems to capture phenomena from the real world (simple click, video, sound, presence, conductivity, etc.) to provide their works with a form of perception and thus invite the spectator to interact with them in real-time. Beyond this feedback loop, it is also possible to make this interactivity more complex by introducing algorithms that allow the system to invent its own reactions and thus allow behaviors to emerge from the work that are not anticipated by the author or spectator.

For interactivity to happen, it presupposes that the spectators perceive themselves, in contact with the work, this possibility of feedback and are tempted by dialogue. However, in the case of monumental projection mappings, this is not self-evident. In this type of device, the crowd – whose presence is sometimes fortuitous – quickly places itself in an immobile posture of contemplation and it then appears difficult to provoke the desire for intervention. In addition, it is difficult to make interactivity visible. The spectators do not necessarily make the step of splitting the crowd in search of an interface. They just look up at the show. How can we ensure that there is interactivity? The following three creative examples provide an overview of how interactivity translates into monumental projection mapping today.

In 2016, as part of *Cœur de ville en lumières*, the façade of Saint-Pierre Cathedral in Montpellier hosted an interactive and participative projection mapping entitled "À toi de jouer!" The public walked along a street overlooking the entrance to the building and could contemplate projections in front of them that invaded the facade. At first glance, there was no indication of spectator participation. To grasp the entire system, it was necessary to lean over the low wall bordering the street to see below a barnum in front of which a queue was forming. In turn, the few curious people who had had the opportunity to go down to make the queue entered the arranged tent. Inside, a touch screen offered a series of buttons to choose, from a gallery of predefined effects, the sound and images that make up the projection. Between the lack of visibility of the cabin and the discouraging queue, the existence of interactivity could escape the spectator who was then content to watch a classic monumental projection mapping. Through this project, we see the need to think about both interaction design and user experience, but also what we could call scenographic work, paying particular attention to the visibility of the work's "sensitive area".

Every year, the Fabre Museum in Montpellier offers an evening entitled *François Xavier n'est pas couché* during which students invest in the place with creative projects. During the 2017 edition, students were able to create a monumental interactive mapping camera. Three luminous spheres reacted to the spectators' touch to trigger both generations of particles on the facade and an increase in the power of a sound environment around the sphere. The originality of the project lay in its transmedia character, the announcement of the mapping having started well before the projection through a fictional story enriched over the weeks on social networks. The main character, who appeared from the future, talked about the museums of his time and invited everyone to attend the inauguration of this interactive installation of the Fabre Museum, the only place that has remained as it is throughout time. Such a form of upstream "mediation" does not completely reduce the *in situ* difficulty of making interactivity visible. However, it was interesting to observe the small groups forming spontaneously around the luminous spheres, taking advantage of the more localized and palpable interactivity associated with sound.

For its part, the TeamLab collective proposed an immersive projection mapping exhibition at the Grande Halle de la Villette in 2018. The spectators are surrounded by 360 degrees of projected universes that evolve in contact with the gestures and movements of the spectator. However, once again, the number of spectators has grown so large that the interactivity is complex to grasp. Some manage to find interactivity where there is none. We see that

Points of View: Sound, Projection and Interaction   155

this interactivity can extend into the spectator's imagination. This echoes other experiences, such as those of students who anticipated the possible dysfunction of their interactive device and who randomly trigger events in the work. Despite a lack of interactivity, some spectators still imagine that they have generated a visual with their movements. On the other hand, others give up completely with the desire to identify their influence within the work, wander around the space and contemplate the creations: interactivity becomes secondary.

To help interactivity to take place, one can consider collecting a mass of data with cameras or presence sensors, but how can we translate them visually so that the audience can finely grasp the relationship woven with the projection mapping work? At this level, using terminals appears to be an easy solution. The general public quickly understands their use (pressing a button, moving their hand over a sensor, etc.) and then the public/video link is created more easily, while finally restricting participants. To encourage this participation, one could imagine that mediators could guide the public on site, or else think of a route that would lead spectators past the terminals, but it would be a matter of reconciling movement and contemplation. Perhaps more generally, it would be a matter of accustoming the public to this possibility of interaction or diversifying their experience in mapping. While Montpellier residents are now used to an event such as *Cœur de ville en lumières* organized every year, the city is above all committed to making the event a sound and light show, a family and friendly evening without necessarily wanting to develop unusual or reflective experiences, for example on the relationship of the city dweller to heritageor public space.

9.3.2. *Small interactivity and projection mapping*

Smaller projection mapping works seem better suited to interactivity with the public, for example street mapping. During the Video Mapping Festival in Lille, in 2018, Olivier Cavalié's creation "Infini prospect", projected against a brick wall on rue de Béthune, offered an interactivity that worked, in the sense that a crowd of spectators had really engaged in the work's animation. In front of a *Kinect* on the ground, the audience staged themselves to bring the visual to life. Something happened immediately: the exchange took place in a kind of communion at work. For example, teenagers became aware that they were gesturing in the middle of a crowd that watched them interact with the work. Their behavior changed in response to both the interaction with the work and the desire to perform. In this dual situation, these young people were having fun with the interactive device while fully contributing to it.

Even if the dialogue with the work does not directly involve the entire crowd, it seems possible to create an effect of empathy with the interacting spectator and to spread the sharing of emotions to the surrounding audience. In this respect, a projection mapping projected in a smaller place, intended for a smaller crowd, allows us to create a link and to release a strong intensity.

9.3.3. *The future of interactivity in projection mapping*

Interactivity has long been used by video games and is deployed in homes by diversifying experiences, involving the body, as with cheap *kinects*. *In the* same way, projection mapping can see an interest in engaging the spectator in a relationship with the medium and diversifying experiences through new technologies. In the case of projection mapping, two conditions seem to favor the interactive experience in particular.

On the one hand, the size of projection mapping seems to have to be considered closely. With monumental projection mapping, the public has more difficulty perceiving interactive devices and more generally its relationship to the work. This issue should be further explored. On the other hand, a more localized experience, involving a small crowd of spectators, facilitates the intelligibility of interactivity, the relationship to the work and its sharing.

On the other hand, projection mapping, whether monumental or smaller, interactive or linear, gives a very specific place to the medium. The surface of the "screen" can be varied, colored, with levels of reflection, transparency, roughness, offering a wide range of perceptions that can contribute to the narrative. Interactivity can therefore make it possible to consider the medium as a sensitive material. Beyond interactivity, the use of projection mapping should be considered in all its specificity as a link between content and support. During the Video Mapping Festival in Lille, in 2018, Gervaise Duchaussoy's creation "Boîtes en dialogue" in the rue de Béthune illustrated this principle with great sensitivity. Two animated characters, a man and a woman, trapped in an EDF painting, communicated with the help of ventilation pipes on the principle of communicating vessels. The author managed to use this unattractive setting during the day to create a nocturnal poetic dialogue, a tale that suddenly offers another image of the urban environment, reflecting a real process of reflection on the capacity of content to give a central role to the medium.

9.4. References

Arregui, M., Casero, S., Maduell, E., Vilanova, S. (2011). Kernel festival, Telenoika, Desio.

Delavaud, G. (2010). Historique du terme audiovisuel, présentation à Archimages. Université Paris 8, Paris.

Deshays, D. (2006). *Pour une écriture du son*. Klincksieck, Paris.

Deshays, D. (2010). *Entendre le cinéma*. Klincksieck, Paris.

Eisenstein, S., Poudovkine, V., Alexandrov, G. (1928). L'avenir du cinéma sonore et le contrepoint orchestral. *Sovietski Ekran*, manifeste, 32.

Jacquet, Y., Vaquie, T. (2010). *Saint Gervais*. AntiVJ, Geneva.

Pont, J., Fernandez, A. (2014). *Evolucio*. Onionlab, Prague.

Tardy, R., Vaquie, T. (2012). *Omicron*. AntiVJ, Wroclaw.

Us, A., Kolokolnikov, Y. (2018). *Loneliness*. Sila Sveta, Seattle.

Webography

Loneliness, USA (2018):

O (Omicron), Poland:

Onionlab, Evolucio Signal Festival:

St Gervais, Mapping Festival (Switzerland), AntiVJ (2010):

Telenoika Audiovisual Mapping, Kernel Festival, Desio (2011):

PART 3

Production and Dissemination

10

The Factory of the Future, Augmented Reality and Projection Mapping

10.1. Introduction

Projection mapping is not limited to leisure, the entertainment industry or marketing, where it now occupies a recognized and increasingly important place. Projection mapping finds a place in traditional industry. More precisely, the use of projection mapping potentially makes sense in the "factory of the future" approach that is currently emerging in industrialized countries. Technologically related to augmented reality, projection mapping, a major consumer of hardware and software resources in information technology, has real potential for use in digital and connected factories. Demonstrators prove this and, beyond that, complete systems are already operational within the manufacturing facilities in some companies.

10.2. The factory of the future

10.2.1. *The process*

The concept of the factory of the future[1] is entirely linked to the context of globalization and the relocation of a significant part of the manufacturing industries of Western countries (plus Japan and South Korea) to low labor

---

Chapter written by Pascal LEVEL.

[1] See for example (last accessed Feb. 2019), http://industriedufutur.fim.net/ and www.pfa-auto.fr/wp-content/uploads/2016/03/Guide-pratique-Usine-Automobile-du-Futur.pdf.

cost countries. The factory of the future approach is the result of highlighting the importance of maintaining an innovative manufacturing industry in these countries. As far as Europe is concerned, the factory of the future approach is thus part of the 8th Framework Programme for Research and Development (FP8) and its *Horizon 2020* and *Factories of the Future* programmes.

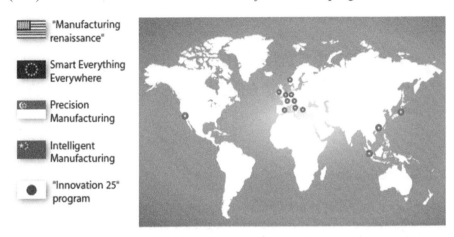

Figure 10.1. *The "factory of the future" approaches in the world. For a color version of the figures in this book see, www.iste.co.uk/schmitt/image.zip*

The factory of the future is the name given in France to this approach to modernizing the production tool in response to transitions that are primarily digital, but also societal or organizational in nature. The development of the plant of the future is based on mature technologies, either in the process of being developed or yet to be defined. The same approach is being developed in several global economic areas under different terms (Figure 10.1). For the Europe zone, according to the member countries, the approaches also have different names (Figure 10.2).

The factory of the future approach concerns all sectors of the company. From design to production, from organization to marketing, it largely integrates the use of digital tools that tend to blur the line between industry and service. However, the factory of the future is not a "simple" production tool. As a link in a global industrial strategy, it is the tool of this strategy. In the concept of the plant of the future, several issues are at stake, such as market developments, technological supply, the societal dimension, the ecological dimension and organizational methods. Projection mapping is clearly part of the "technological offer" challenge. This point will be developed further below. The other issues will not be, and we refer the reader to the sources cited at the end of this chapter to find specific developments on these issues.

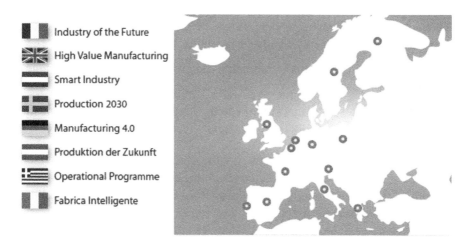

Figure 10.2. *The "factory of the future" approach in Europe*

10.2.2. *The technological challenges of the plant of the future*

The delivery of a technical product whose functionalities are in accordance with the order, within an agreed time and price, is one of the major challenges of this new industry. The personalization of the product, whether it is the response to an individual customer or a company or a community of customers, requires the development of what are nowadays called agile manufacturing processes, that is processes that can be quickly modified, reconfigured, to precisely meet changing demand. In this context, the key concepts are interoperability between machines and the establishment of modes of cooperation between all "actors", including machines and people. The flexibility induced by this need for speed, adaptation and potential quasi-individualization is totally dependent on the development of information and communication technologies which, in a transversal way, allow the integration and connection of the other technologies used. The factory of the future is above all digital and connected.

10.2.3. *A digital and connected factory*

In the industrial context, information and communication technologies support the use of professional software suites also called "software packages". These software packages are used (non-exhaustive list):

– to continuously manage data flows in the production process from design to manufacturing and logistics to maintenance;

– to develop "digital twins", simulating the product, but also all the stages of the production process;

– to implement self-diagnostic methods and, more generally, for the continuous monitoring of tools and products.

The dedicated software packages or software, the development of which has been meteoric in 30 years, now cover all the needs and potential areas of the plant of the future, including, as far as we are concerned, the modelling of products and processes. Through their ability to model, integrate and quickly understand the impact of changes resulting from a desire to constantly adapt products and processes, software packages are the basic building block of the digital and connected factory and are the tool at the service of agile manufacturing that the factory concept of the future implies.

The result of modelling using a software package is very often a "numerical twin" of the object, system or process that is the subject of the study. The very high degree of performance of the software packages means that the quality of the similarity depends more on the resources that have been allocated to the modelling, such as the budget or time allocated, than on the intrinsic quality of the software package. In other words, by putting the means into it, it is possible to generate digital twins of a desired quality level. Once the models have been developed and validated, virtual digital twins or prototypes allow access to an optimization space that is conducive to simulating the effect of any supposed modifications or improvements before any attempt at material realization. This optimization space is virtual in nature. It is basically an algorithmic space where the description of products, processes and their respective behaviors is, in a transparent way, put into equations before they are, also in a transparent way, solved. The numerical twin is therefore a set of equations representative, with a chosen and assumed level of precision, of the behavior of the object or process being studied.

The results are therefore by nature numerical and can easily concern several hundred thousand simultaneous variables today. Their interpretation or even their simple appreciation is beyond the reach of the user and requires the use of a layer of interface software packages linking the digital world to the real world. Progress in sensory restitution, essentially visual for the moment, thus makes it possible, beyond the "simple" simulation of physical and organizational behaviors and the delivery of associated quantitative results, to have digital objects, essentially images, virtual, that can be superimposed on real objects. It is the existence of these digital image objects and their superimposability in the real world that underlies the concept of augmented reality.

10.3. Augmented reality[2]

10.3.1. *Simple definition*

The word "reality" refers to what actually exists and the word "augmented" refers to something that has been made larger, more pertinent, something that has been increased or has grown, often implying that the added parts are of the same nature. Put together, these two words correspond well to the different definitions of augmented reality that can be found when querying a web browser and that can be summarized as "a technique that superimposes an updated digital representation on reality". We will stick to this definition, even if the term augmented reality tends to be questioned because, technically, it is the perception of the user that is augmented and not reality.

To speak of augmented reality is therefore to speak of a set of interfacing techniques, of a computational nature, which allow the superposition of a 2D or 3D model to reality. Although it can concern all sensory domains, augmented reality is still most often used in the visual field or, concretely, it allows us to mix several sequences of images, of which at least one represents reality, for example the video of an industrial device or a machine.

Thus, when it comes to the use of a visual representation of reality, it is important to understand that it is not the latter that is modified, but rather the perception, here visual, that we have of it.

---

2 For a follow-up of the latest news in the field of augmented reality, see the monitoring site in this field: www.augmented-reality.fr/la-veille/.

10.3.2. *Some chronological references for augmented reality*

Augmented reality is now about 60 years old with the presentation in 1962 by Horton Heilig of his "sensorama", a helmet equipped to simulate some scenes of everyday life (Figure 10.3). Heilig's machine is more of a virtual reality, but it is generally accepted that it prefigures the foundations of augmented reality.

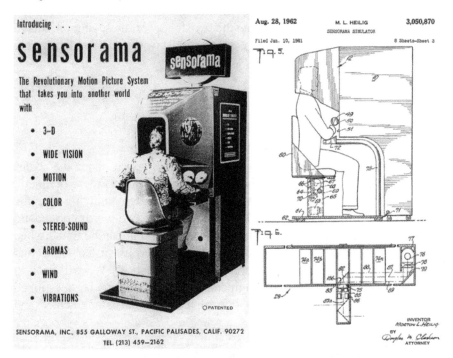

Figure 10.3. *H. Heilig's sensorama*

It was not until 1968 that Ivan Sutherland invented the first transparent vision helmet. It is a so-called head-up display machine that is controlled by the movements of the head according to six degrees of freedom. The first operational realization of this head-up visualization helmet concept was made by Steve Mann who, in 1980, introduced the *Eye Tap*, a device that displays information in real-time in front of the operator's eyes (Figure 10.4).

It was in 1992 that the terminology "augmented reality" appeared, proposed by Tom Caudell and David Mizell of the Boeing company, but the precise definition that still refers today is due in 1994 to Paul Milgram and

Fumio Kishino[3] who defined precisely the limits of the various subparts of "mixed reality", from pure reality to pure virtuality, including augmented virtuality and, of course, augmented reality. This proposal is summarized by the Milgram continuum frequently found in augmented reality sources (Figure 10.5).

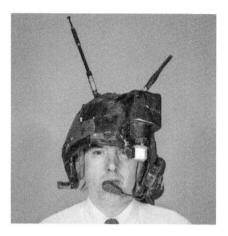

Figure 10.4. *Steve Mann and the Eye Tap (source: Steve Mann Photo: Richard Howard; Time Life Pictures; Getty Images)*

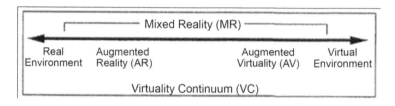

Figure 10.5. *Milgram Continuum*

Finally, modern augmented reality, in the sense that we know it today, appeared in 2003 with the mobile game "Mozzies" developed for Siemens mobile phones and whose goal was to capture virtual mosquitoes evolving in our real environment. The computer graphics[4] (Figure 10.6) proposed by Manon Boschard and Lucas Zapata usefully complement these historical landmarks.

---

3 See www.alice.id.tue.nl/references/milgram-kishino-1994.pdf.
4 Computer graphics available in several web sources including: www.anthedesign.fr/autour-du-web/realite-augmentee-ra/.

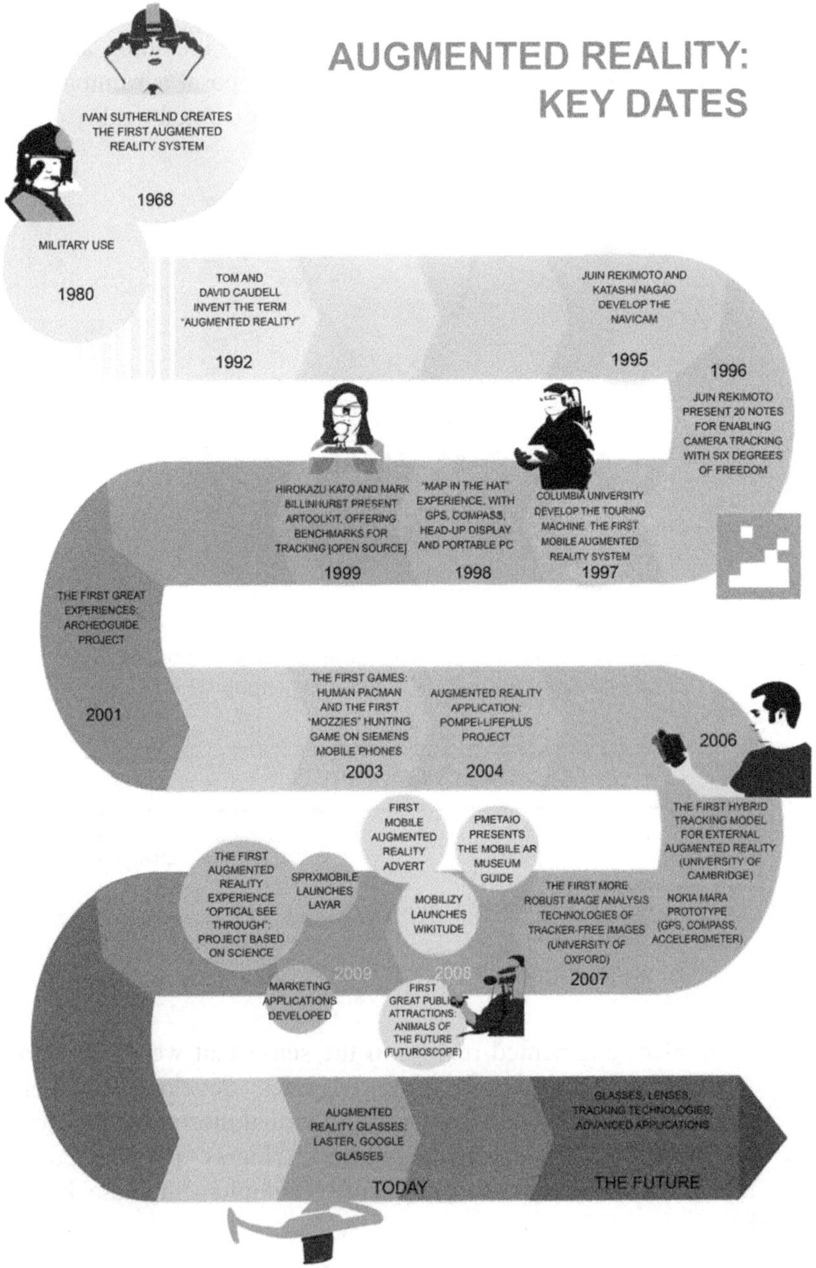

Figure 10.6. *Some key dates in augmented reality (source: computer graphics by Manon Boschard and Lucas Zapata)*

10.4. Factory of the future and augmented reality

Augmented reality technology, whose beginnings date back some sixty years, has developed in parallel with the development of computer computing power, constantly seeking to exploit its full potential. The details of these technologies are not specifically developed in this chapter. It is simply important to note that thanks to the fantastic computing power of today's computer machines, augmented reality technologies are now able to offer extremely realistic real-time effects that are sought by the gaming industry, in education, in the medical field and also in *manufacturing*.

In this field, in the heart of the plant of the future, real-time is very useful when augmented reality is used as a tool for precision comparison. The aim of the tool is to enable virtual data to be compared very quickly with reality. Here, the virtual data corresponds to the goal, to what you want to achieve, to what is planned. It is important to constantly remember that these virtual data result from a modelling process whose parameters are induced by the needs formalized by the plant concept of the future. As for reality, at a given moment, it is composed here of the process or product in its state of realization at that moment. This comparison tool, this computer-based gauge, is a key tool in terms of guaranteeing compliance with specifications, the diversity and high level of requirements of which are at the very heart of the concept of the plant of the future. Having such a comparison tool is fundamental. In the factory of the future, all areas of production are potentially affected.

As far as the product is concerned, it is the complete cycle that may require, at each stage, objective/state of the place confrontations. Chronologically, it goes from the conception and design to the manufacture of the parts and their assembly to the maintenance of the finished system in its operating cycle. For the production tool, all the equipment, its implementation, its control, robotics are potentially users of real-time reality/objective comparison methods.

Measuring a deviation, detecting a dispersion, observing a gap, an omission or an error are all elements taken into account in quality and specification compliance procedures. *In input*, augmented reality technologies provide objective data for these control processes. The purpose of the latter, beyond the launch of an alert of any kind, is to help in diagnosing and proposing remedies in relation to the failures observed. As an *output of* the continuous control cycle, augmented reality is widely used to project instructions into the real environment. These instructions may correspond to useful data during normal use, but also to information useful to remedy a failure situation.

10.5. Augmented reality and projection mapping

As it has been written earlier, to speak of augmented reality is to speak of a set of computer-based interfacing techniques that allow the superposition of a 2D or 3D model to reality. Obviously, this definition, although it may be reductive, can be applied to projection mapping as an image or video projection technology. In its use, projection mapping actually mixes real and virtual images. However, the two approaches differ quite clearly if we look at them through the *package* of technology they require respectively.

Projection mapping is used to project light or videos onto any surface and therefore, in fact, uses these surfaces as support screens. The surfaces, support screens, unambiguously hold here the role of reality in the process of superimposing a 2D or 3D model that is common to both technological groups. In their most common acceptance, augmented reality techniques require the use of a common medium for displaying both the 2D or 3D model and reality. This can be, for example, the screen of a smartphone or tablet or that of a headset or specific glasses. Reality and model each play the role of a layer. There is no video projection. Technically speaking, the model layer, which is virtual by nature, is only visible by displaying it on the screen. The reality layer results from the capture of an image or video that is then calibrated and "placed" on the model layer, unless the actual image was captured before the model layer was applied to it. In both cases, the virtual nature of the model allows the user to choose the degree of transparency of the image to enable "observing" the differences by superimposing the two layers.

In projection mapping, the "screen" is not the support of the superposition of two layers. It is one of the two "layers" and therefore participates in images to be superimposed and mixed. It is, by its shape and patterns, the real image on which the model layer is projected. It can, of course, have its own transparency as in the case of a projection on a veil carrying or not a pattern. However, this does not change its role as a receiver layer. Although part of the same technological approach as augmented reality, it is clear that the development of projection mapping and its practice require specific digital tools and techniques. However, just as the techniques specific to augmented reality have not been developed in this chapter, those relating to projection mapping will be developed here.

10.6. Future plant and projection mapping

10.6.1. *Some preliminary considerations*

Augmented reality and projection mapping are brother and sister and children of the same technological developments. Thus, like augmented reality, projection mapping makes full use of the computing power of today's computer machines to offer extremely realistic real-time effects. Like augmented reality and because it offers similar potentialities, projection mapping is also beginning to be implemented in manufacturing.

Before presenting some projection mapping applications currently being implemented in the industry, it is important to highlight some common points and significant differences in the industrial deployment of the two technologies.

Augmented reality requires the use of an intermediate medium to support the display. The operator who uses an augmented reality process is therefore equipped with specific equipment. He holds it in his hand or puts it on a stand if it is a smartphone or a tablet. It can be worn on the skull, face or wrist for specific equipment. There is therefore a potential constraint related to the wearing or handling of this equipment.

On the other hand, since the equipment is, by nature, relatively small and light, the operator has a real possibility to move and choose the position of use that suits him best. In this case, the computing power of the computers is used to calibrate and adjust the two "layers" in real-time. If the operator or part of his body enters the field of the real image, his image is displayed on the screen and he is therefore warned.

Projection mapping requires the use of a model projection device. Although there is a constant trend towards miniaturization, this device occupies a certain place and must be protected against the risk of misadjustment, breakage or injury in the event of a collision with the operator. If the operator or part of his body enters the projection field, a more or less important part of the useful information is no longer available on the projection surface since it is projected on the operator. These considerations lead to different manufacturing applications. Both technologies are part of the toolbox of the factory of the future, which is a digital and connected factory. Both are immersive and interactive in nature. They are vehicles for communication between machines and/or between operators and machines. In various forms, they disseminate data from the company's information system and in return collect and transmit data to the same system. They have the strength and modernism, but also the weaknesses because they are unable to function autonomously and extremely sensitive to both internal IT risk and cyber-risk.

10.6.2. *Some examples of projection mapping in manufacturing*

10.6.2.1. *Methodology point*

Industrial methods and more particularly manufacturing methods are at the heart of the company. They are generally well protected and therefore poorly publicized. The few examples highlighted here are the result of a web search. The main keywords or key concepts that have been used to detect uses of projection mapping in industries are few and are cited below: augmented reality, Manufacturing Augmented Reality Support, DIOTA (start-up active in the field of augmented reality), factory of the future, industry 4.0, factory of the future. A keyword search combining projection mapping and industry does not produce results (February 2019). Projection mapping is not identified as such, but seen as an augmented reality technique. The requests concerned video resources available on the Web. They were viewed until they found or not sequences using projection mapping technology.

10.6.2.2. *Projection mapping and assembly*

It is the field of assembly of parts or sub-assemblies that seems to be the most developed application sector today for projection mapping in the factory of the future. In this field, manufacturing devices exploit the ability to compare in real-time the reality, that is the system being assembled, and the model, that is the next assembly step as such or the result of the next assembly step. As described in section 10.4, the assembly station, of which the projection mapping device is an integral part, allows virtual data to be compared very quickly with reality. Virtual data corresponds to the purpose, what you want to achieve, what is planned, what is compliant. The reality here can be the system composed of the object being assembled, the assembly tools and the operator in charge of this assembly.

10.6.2.2.1. Case of the assembly of a motorcycle engine[5]

This involves assembling a chain guard on an engine block after completing the tightening of two-timing belt tension pulleys. The video perfectly illustrates the use of a projection mapping device integrated into an assembly station.

Figure 10.7. *Three views of a video mapping assembly device for a motorcycle engine*

---

5 See "Industry 4.0: Augmented reality system for production", www.youtube.com/watch?v=0m67O1Em7dY.

Although not specifically indicated, it is nevertheless likely that this video will present a demonstrator designed to show a range of possibilities and not an assembly station in an industrial situation. The complete device is not described and, in particular, the error diagnosis device is not indicated.

The projection mapping tool is used for several tasks. One of the tasks is the projection on the assembly workbench of data relating to the operation to be carried out. Thus, for the pulley tightening operation, we can read a sequence of successive messages *take the wrench, please go on the first screw, tighten at 15Nm, Torque 15.23 Nm Angle: 0.1°* which ends with *Position wrench in cradle*. The data here correspond to instructions (*take the wrench*) or diagnostics (*Torque 15.23 Nm Angle: 0.1°*).

Another task is the projection, directly on the engine during assembly, of aids for correct assembly. For example, for the pulley clamping operation, a green spot of light is projected on the first pulley to be clamped. For the assembly of the housing, it is successively the housing that is pointed by a flashing white light spot, then when the operator has grasped it, it is the location where it must be placed that is illuminated in the same way. In both cases, it is the projection of data corresponding to instructions.

The third task identified concerns the projection of alerts or diagnoses during assembly. For example, an orange spot of light is projected over the entire engine surface when the operator tries to tighten the wrong pulley or an orange spot of light is projected on an unsuitable screw container when the operator extends his hand towards it.

10.6.2.2.2. Case of the interior trim of an airframe

In February 2019, several videos relating to the STELIA Aerospace company's website in Méaultes and its collaboration with the start-up DIOTA were posted on the web (see webographic sources). On these promotional and demonstration videos and particularly on one of them[6], we can, between the times 3'08" and 3'17", see an operator position a dressing bracket on the inside face of an airframe using a "virtual" mounting template projected on the surface. It seems that the virtual template gives at least two instructions – the place and the order of assembly – since only the correct location is illuminated at the *appropriate* time.

---

6 See "Bienvenue dans l'usine du futur, robotisée et digitale de STELIA Aerospace", www.youtube.com/watch?v=5PBY89AzhrI.

Unlike the example of engine mounting described in section 10.6.2.2.1, this is an effective industrial system. The commentary on the other videos in the same series tells us that this method has totally replaced the one in which the journeymen editors operated from complex assembly plans and complicated handling.

Figure 10.8. *Airframe assembly video mapping*

10.7. Conclusion

Like augmented reality, projection mapping results from the juxtaposition and simultaneous or non-simultaneous use of a large number of sophisticated technological building blocks that continuously benefit from the development of information and communication technologies. The factory of the future, which is digital and connected by nature, offers a suitable field for development and application beyond the simple projection of promotional or extension products. The production tool itself, manufacturing, is beginning to integrate projection mapping devices, just like a conventional machine or robot. For the moment, it seems that it is the ability to superimpose and compare in real-time two states, one of which is the real product being developed, that attracts the attention of manufacturers. Operational assembly devices that use projection mapping in this sense have been developed. The support object of the assembly is the screen on which virtual assembly templates, alerts and advice are projected. These are probably not the only applications that exist and they will certainly only grow.

Webography

Demonstration of the Lynx system, Stelia Aerospace:

Diota:

11

Heritage Mediation through Projection Mapping

11.1. Introduction

Heritage is a cultural element whose values attributed to it by a community have transformed its status. Monuments have been of cultural, historical and symbolic interest for centuries. The Seven Wonders of the World, erected as icons by the Ancients, represented for them the most perfect works ever made; the only survivor, the Pyramid of Giza, attracts millions of tourists annually. The remains of antiquity, which served as evidence of Greek and Roman writings among humanists, then represented a documentary value for scholars described as "antique dealers" (Choay 1999) until the 19th Century. The historical monument, later consecrated by other values, such as cognitive, then becomes central in the discipline of art history.

From objects of contemplation and study, monuments have evolved into objects of artistic creation through different practices, for example, by shedding light on them from the 19th Century onwards in Paris, or by interventions such as Christo and Jeanne-Claude's packaging on icons such as the Reichstag or the Pont-Neuf. But monuments, while serving as a medium for artistic practice, are also sometimes vehicles for social practice, as in Krzysztof Wodiczko's projections, which offer a critique of the world and inequalities. The built heritage represents rich opportunities for artistic creation thanks to its status which distinguishes it from other common

---

Chapter written by Alexandra GEORGESCU PAQUIN.

buildings. This monumental status makes it possible, on the one hand, to offer a rich and complex basis for an artistic intervention, which can draw on the memory of the place or bring out particular forms. On the other hand, heritage represents a showcase for artistic creation, thus enriching the city's experience.

In this context, projection mapping has naturally been applied to heritage buildings, both for the interest of their projection surface and for their iconic status. Whether during festivals (such as festivals of light, digital creation or projection mapping), special heritage days (Nuit Blanche (Sleepless Night), Heritage Day, etc.), to inaugurate cultural spaces or commemorate events, or to project promotional campaigns or advertising content, temporary interventions through projection mapping are numerous and heritage offers a suitable space to encourage significant distribution. The heritage that serves as a support is often the monumental, or iconic heritage, in particular cathedrals (Amiens, Chartre, Reims, Strasbourg, Notre-Dame, or elsewhere in Santiago de Compostela, the Italian *Duomi*, etc.). The monuments on UNESCO's World Heritage List, such as the Taj Mahal, the Eiffel Tower, the Sydney Opera House and the Great Wall of China, among others, also serve as an anchor for a show that enhances the visitor experience. But what do projection mappings bring in return for this already monumental heritage? Other interventions, though less numerous, are designed as a permanent installation, and knowingly transform the visitor experience, whether it is for a monument or a museum.

This projection technique is applied as much onto and inside the heritage. It produces content that can be applied to different registers and spatial contexts, depending on the relationship that the content of the projection has with the heritage that supports it. On the one hand, the façades already present themselves as a space of communication between the building and its urban context, which can be enhanced by an architectural projection. On the other hand, the interior space is increasingly invested in by projection mappings, both for the enhancement of the space itself and for the enhancement of a heritage that is exposed there, as in the case of museums.

Heritage is a complex medium, already carrying meanings through its symbolic component. Projection mapping, by affixing a media layer to it, reinforces the monumentalization of heritage, in the sense that it carries a heritage discourse beyond the built environment to transcend it in an interactive collective experience:

"The projection has some forerunners in theatre performances, baroque festivities or sacral buildings but the existence of contemporary projection methods marks a decisive step as the surface of architecture became permanently changeable and a means of communication that goes beyond the 'symbolic' way of communicating which has always been a part of the perception of architecture." (Krautsack 2011)

In this sense, projection mapping serves as a spatial augmented reality tool, as its academic name suggests (Raskar *et al.* 1998). This double communication must be taken into account when designing the mapping, to avoid antinomic overlaps.

This text aims to deepen the reflection on the link between projection mapping and heritage in its various spheres of presentation, in order to identify the specificities and characteristics of this association. First, the symbolic status of heritage and the consequences of its actualization will be presented in order to understand the implications of the different mediations. Then, we propose sketching a typology of mediations at work between the heritage as a support and the architectural projection: transcending, combined and self-reflective.

11.2. The symbolic value of heritage

Before grasping the implications of adding a layer of digital meanings to heritage, it is important to reposition the concept of heritage. Instead of the essentialist conceptions of heritage, it is considered here as a social construction. It is a cultural good (in the case of tangible heritage) that has gone through a process of "heritagization", or heritage-making, transforming its status as a cultural element to that of a heritage element. Heritage can be certified by experts according to pre-established criteria, but it can also be considered by the assignation of its related values by a community. These values may be, for example, historical, aesthetic, social, positional, material or use-oriented. A selection is thus made between all the material cultural traces, giving some of them a special, symbolic status.

This process of patrimonialization is certainly not static. Being integrated into the heritage ecosystem (Morrisset 2009), it is dependent on the surrounding contexts. Since these contexts are changing, a heritage element may change this status as a result of the lack of recognition or the loss of the values attributed to it.

First confined to the notion of monument, the concept of heritage has gradually undergone a "a threefold extension: chronological, typological and geographical" (Choay 1999). Since the 1980s, the phenomenon described as "heritage inflation", in which everything could be considered as heritage, has been considered to have led to a loss of meaning because of the proliferation of heritages.

However, creative interventions can constitute a strategy to enhance the heritage through contemporary lenses (Georgescu Paquin 2014) by resemantizing it. Involving heritage in a dialogue with a contemporary work that intervenes in its memory strata or in its aesthetic forms is one of the strategies to confront the meaning that this heritage has for a society at a given moment. This "extraordinary" status of heritage, in the sense that it is detached from other usual and cultural elements, is what makes it attractive as a projection medium. It is therefore not insignificant or neutral, but this symbolic charge is not always taken into account when creating the visual and narrative material projected therein.

The dialogue or contrast between the old and the new is often a source of polemics and contradictory visions; some consider heritage as a project, others as an object to be preserved, untouchable. Unlike the integration of contemporary architecture into historic monuments, which often give rise to strong opposition, projection mapping generally does not represent a threat to the heritage. As long as heritage is not exploited solely for advertising and commercial purposes, "heritage emotions", to use the words of ethnologist Daniel Fabre (2013), are aroused by the positive experience generated by architectural projection, and not by the controversies that could be generated around the juxtaposition of old and new. Due to its ephemeral and normally non-intrusive nature, projection mapping is a sustainable solution for contemporary creations in the heritage (Salas-Acosta 2014).

11.3. Projection mapping as a means of cultural heritage mediation

Projection mapping can represent a mediation channel by enriching the meanings of its support or context. Mediation is considered as a space of communication; not only an intermediary, but also where discourses belonging to different universes can be constructed in this system: "an action involving a transformation of the situation or the communicational device, and not a simple interaction between elements already constituted, and even less a circulation from one element from one pole to another" (Davallon 2004, p. 43).

Projection mapping, by affixing a narrative or artistic discourse to a heritage place or object, which is already significant, opens up a virtual space where several interpretations and representations can be negotiated, which is what mediatecture defines: "Mediatecture is the orchestration and temporalization of space, the loading of meaning into space as well as the creation of a sphere of communication" (Kronhagel 2010, p. 3). The façade is the most commonly used surface for projecting video mapping works, but the interiors and objects also constitute a relevant space to reinvest a heritage.

Projection mapping, as a means of mediating the heritage, can be effective in various ways: educational, creative, evocative or emotive. The projection mapping technique allows different registers of mediation, depending on the intention and type of content projected. The following typology proposal represents different interactions between heritage and projection mapping, creating three main types of mediation: in transcending, combined and self-reflective.

11.3.1. *Transcending mediation*

Transcending mediation refers to when heritage is used as a canvas to projects some content that is unrelated to the meanings of the place, for example to convey a message. In this case, projection mapping is not created specifically to enhance the building, but to add a different message, either as a promotional tool or to promote another heritage. This type of mediation is therefore considered to be transcending. For its centrifugal force, this type of mediation transcends the heritage upon which the video mapping is projected. Therefore, heritage is the basis for the diffusion of an external discourse. This type of mediation ensures the visibility of heritage with the possibility of enhancing its forms. However, this is not its primary objective. Thanks to its monumental importance, both physical and symbolic (through its collective recognition), heritage becomes a place of gathering and collective experience, around the projection event.

11.3.1.1. *Promotion*

It is precisely because of its capacity to bring people together that heritage is sometimes chosen for commercial purposes under the pretext of entertainment. However, the concept of mediation is debatable according to the degree of connection to heritage. In 2011, Bacardi projected a mapping on the Kursalon building in Vienna, a key place in the Viennese musical world since the 19th Century, to promote its "Bacardi Together" campaign. The project focused on elements related to rum culture, following the prerequisite of incorporating Cuban heritage, among other things, rather than making reference to the history of the place. The same year, Les Garçons designed an immersive installation with Les Vandales to launch Dior's *J'Adore* eau de parfum during an evening at the Baccarat Museum in Paris.

Other brands use the built heritage as a screen that allows them to shine and support their message, as in the case of Coca Cola, which has worked with the Bulgarian agency MP-studio to promote its brand. For example, to celebrate *Culture Night* in 2016, they projected an animation on the Axelborg building, a Danish built heritage site, to actually promote the new Coca Cola Zero line.

This type of association does not result in a mediation process; the building is only a vehicle for dissemination, but does not enrich it. The art historian Santos M. Mateos Rusillo, who specializes in communication and cultural heritage, described two situations from a critical perspective in his *Miradas desde la copa* platform. The first was the TATGranada (*Talking about Twitter*) event, which was advertised in 2013 by the projection of the company's famous bird on one of the Alhambra towers; the other was a projection mapping to promote the new products of the BMW brand, this time on the Seu Vella castle tower (aspiring to be included on UNESCO's World Heritage List), with the permission of the administration of the city of Lleida (Catalonia). Wheels, cars, as well as the company logo were projected there, culminating in the *in situ* unveiling of the car, a content totally detached from the heritage.

The association between art and advertising is not always denounceable, and can contribute to the dissemination of what is considered to be elitist art: "While the works of art that are shown in advertisements are often already acknowledged as 'high art' by experts, it is their appearance in advertisements that gives them the final signature from a wider public" (Vermehren 1997; cited in Hetsroni and Tukachinsky 2005, p. 94). However, when it is the brand that is projected on the heritage, it monopolizes the message and reduces the medium to a distribution channel for purely commercial content. The iconic heritage serves to endorse, in a way, the message disseminated, by giving it a more noble patina, such as the way neoclassical and romantic styles are usually used as an asset in advertising to bring good reputation and good taste (Hetsroni and Tukachinsky 2005, p. 103).

Another type of promotion, this time used in the tourism sector, uses projection mapping on mainly institutional buildings to celebrate the "brand" of a city or region. For example, in spring 2018, the Barcelona Provincial Government building, designed by the famous modernist architect Puig i Cadafalch, was used as one of the vehicles to disseminate the *Barcelona és molt més* (*Barcelona is much more*) tourist promotion campaign. The message of the mapping was clear, by using phrases such as "come and visit the different regions of Barcelona" or by writing the names of the events promoted as a complement to the images, which also refer to traditions such as human pyramids (*castells*) or gastronomy. In this case, the choice of the building places the modernist monument at the heart of a complex that is emblematic for the public institution, acting both as a support and as an actor in the diffusion. Two years earlier, Marca España's *Spain Today* projection mapping claimed to "spread Spanish values" by projecting the video on a sculpture in the form of the letter "ñ". Strategically placed in Potsdamer Platz in Berlin and Columbus Square in Madrid, the projected promotional message was totally detached from its context or support, without any architectural interrelation. The heritage context is not taken into consideration in the projection, in order to focus on the elements that are considered as representative of Spain, and particularly of Madrid through its "talents": its main museums, innovation, sports clubs, gastronomy, events, among others.

11.3.1.2. *Commemorations and historical content*

Though different from the built heritage where the mapping is projected, its message is not only restricted to commercial or promotional purpose; historical narrative content is another case where heritage transcends the message outside its walls. Commemorative acts, for example, are often associated with public institutional buildings such as town halls, which are highly symbolic and identity-based places. Around 2010, in order to commemorate the Latin American countries' independence, these types of monuments were projected upon with a rather didactic approach: "The symbolic role of monuments is completed by the inclusion of images of historical documents, photographs and paintings very similar, in fact, to illustrations that can be found in school books" (Alonso and Romay 2013). This use of historical documents to tell the Great History is also reflected in Sila Sveta's Stalingrad project, which in 2013 commemorated the 70th anniversary of the Soviet victory during the WWII Battle of Stalingrad, projected on the facade of a hotel in the city's central square (now Volgograd). *L'arbre de la memòria (The Memory Tree)*, projected in 2018 to commemorate the 80th anniversary of the bombing of Barcelona during the Civil War, is rather poetic and evocative, straddling artistic mediation. However, its commemorative purpose dominates and, even in an abstract way, it conveys a clear message, among other things by being part of a set of educative and commemorative activities organized by the Barcelona Museum of History as part of this anniversary.

External narrative content can also pay tribute to characters, dead or alive. However, these patriotic demonstrations, as well as these tributes, can quickly slip into the kitsch aesthetic, as in the case of the projection mapping *The Elegant Lady* projected on Bangor Castle in Northern Ireland in honor of Queen Elizabeth II's Diamond Jubilee in 2012, combining historical speeches and rock music (Hanson 2014). More poetic, the projection mapping *Movimientos Granados*, directed by Xavi Bové and projected in 2016 on Barcelona's famous Pedrera to commemorate the centenary of the death of composer Enric Granados, interacted between the movements of a conductor and the music played live in the balconies of the monument, then interpreted in the movement of light.

Finally, the stories of the city are also a source of narrative external to a monument on which it is projected. However, the role of the building here becomes much more central, because it belongs to this history, from which history shines. The heritage chosen on this occasion sometimes differs from the more institutional or expert heritage. For example, during the summers of 2008 to 2013, Quebec City projected a monumental projection mapping to commemorate the city's 400th anniversary with a "historical-poetic" work with the help of a historian and museologist to balance aesthetics with documentary rigor (Georgescu Paquin 2015). The medium, in this case, was a modern heritage, a huge industrial structure, which earned the projection an entry into the *Guinness Book of Records* in 2008. The eagerness for the monumental format was also illustrated for the celebrations of the 400th anniversary of Madrid, in 2017. Óscar Testón's projection mapping, projected in 360° on the four facades of the Plaza Mayor, represented the largest projection mapping in Spain at that time.

One project stands out from this type of mediation: *Cité Mémoire* (Figure 11.1), a Montréal en Histoires project, created by Michel Lemieux and Victor Pilon in collaboration with playwright Michel Marc Bouchard. Deployed on 25 "tableaux" in the tourist areas of Old Montreal, the Old Port and two emblematic hotels, on monumental façades, alleyways and even trees, the city becomes the support for testimonies, often embodied by characters (illustrious or not), of Montreal's history. The aim here is to use the different heritages, small or emblematic, to weave a fragmented and theatrical narrative network, while enhancing the territory through the open-air route. This is completed with a mobile application. Winner of several awards, the concept is planned to be transposed in Paris in 2021, with the collaboration of the writer Alexandre Jardin.

This type of mediation, whose support is detached from the projected content, still makes it possible to rediscover other heritages or to add messages to them that, by their symbolic values, are compatible with the use of heritage. However, the degrees of compatibility are questionable depending on the use of the heritage for promotional purposes.

Figure 11.1. *Cité Mémoire, a Montreal project. Various projections illuminate the urban heritage, public spaces and even trees (source: Montreal en Histoires; photo: Jean-François Gratton). For a color version of the figures in this book see, www.iste.co.uk/schmitt/image.zip*

11.3.2. *Combined mediation*

In this type of mediation, the tradition of VJing (which uses visual material to create real-time performances), is felt. The 3D projection participates above all in the renewal of the experience of a place. When this place is a heritage site, one of the strategies is to pay tribute to its forms or the sources of inspiration of its creation, but not in a didactic way. The final work, therefore, is a fusion between the creation of the video mapping and the place for which it was specifically created, generating a combined mediation. More evocative than descriptive, the projections generally display abstract forms.

11.3.2.1. *Artistic*

Most of the video mapping works projected on architectural facades belong to this category. Playing with illusionary effects, such as construction/deconstruction, appearance of elements, transformation, fragmentation of materials, or highlighting the structural elements of the façade, they are mostly projected in temporary contexts, such as festivals, but also in permanent or semi-permanent installations.

The installation *O (Omicron) of* the AntiVJ collective (2012) is an example of these type of interventions. Inside the Hala Stulecia building (Wroclaw Centennial Hall), listed as a UNESCO World Heritage Site since 2006, this projection mapping focuses on the dome through minimalist visual aesthetics, reaffirming its importance. The size of the building is monumental, accommodating up to 10,000 people. Its 23-meter ribbed dome was, at the time of its construction, the largest reinforced concrete dome in the world. The projection mapping pays tribute not only to the forms of the place, but also to its representation as an innovative place. Enhancing its initial plans and the fact that it represents pioneering architecture as well as a precursor to the beginning of modern architecture and engineering, projection mapping also evokes a futuristic aesthetic.

Another dome was transformed by projection mapping, this time on the occasion of the centenary of the death of Gaudí's patron Eusebi Güell: *Parabolic Gaudí* (Figure 11.2), a creation of Playmodes projected in the central dome of the Palau Güell in Barcelona in 2018. In this case, projection mapping punctuates a guided tour, but in a more artistic and evocative spirit, offering an alternative experience of the space. Located in the central hall, where the frequent organ concerts were amplified by the dome, the projection therefore takes up the essential function of the place and uses the architectural elements of the dome. The hexagons help to create geometric and abstract forms, thus evoking a fundamental aspect of Gaudí's work.

Figure 11.2. *Parabolic Gaudí at Palau Güell, 2018 (source: Playmodes Studio)*

In Montreal, another kind of enhancement takes place at Notre-Dame Basilica; the immersive experience, created by Moment Factory Studios under the name of *Aura*, is an explosion of effects, light, lasers, following a framework that evokes mainly the elements of nature. The studio, which specializes in multimedia events including major concerts (Madonna, Arcade Fire, etc.), offers a style that contrasts with the spirit of meditation or contemplation of churches. While grandiose in its emphasis on magnificent structures such as arches, columns or rosettes, the projection mapping fills the church with an overload of textures and images projected in a rhythm of high intensity without giving the spectator any respite, leaving no one indifferent.

11.3.2.2. *Event-driven*

Several museum and heritage facilities are increasingly using projection mapping works to celebrate an inauguration, reopening or new exhibition, especially in France. It acts therefore as a self-referential promotion of the place, either by evoking its history and representations, or by evoking its content. The objective differs, in this sense, from the following type of mediation, which has been named "self-reflective" and has a more didactic or educational purpose.

For example, two video mappings were projected on the Casa Batlló, a work by Antoni Gaudí, on the theme of the *Awakening of the Dragon* on two occasions. First in 2012 to celebrate 10 years of public outreach. Then, three years later, to celebrate a decade of inscription on UNESCO's World Heritage List by being more interactive with the public. In that case, the narrative is clear; it makes reference to the various interpretations of the building, in addition to representing the architect in his creative process.

More poetic, Atelier Athem has designed several projection mappings on museums or cultural institutions to mark events, such as the Louvre, the Grand Palais, the Palais de Tokyo and the Institut du Monde Arabe. Between promotion and mediation, external projection mapping appeals to people who are not yet in the "sacred" space of the museum. It attracts the eye of passers-by, but also makes the collections look different, almost as an external museum complement. For example, to present the exhibition "Gauguin the Alchemist" in 2017, the projection mapping paid a tribute to the painter by evoking his general work through motifs, topics and techniques. Instead of exposing them too demonstratively, the projection mapping deconstructed the works from the photographic agency of the RMN-GP. On the occasion of the launch of the exhibition "Being modern: MOMA in Paris" at the Louis Vuitton Foundation, the projection mapping on Frank Gehry's iconic building gave rise to reinterpretations of the works displayed inside by adding a dynamic dimension to them, deconstructing them and rebuilding them. Earlier that year, for the inauguration of the Abu Dhabi Louvre, the Pyramid in Paris was chosen as a canvas not only to present the works of the new museum, but also its architecture. The projection mapping included the Parisians in this event whose physical distance with United Arab Emirates was reduced by the direct link established between the two institutions.

Combined mediation therefore refers to the self-celebration of a place or content, through art forms or by marking an event. The junction between the projection and the place, directly united by the semantic layer created in this fusion, makes it an intermediate mediation, which is neither detached from nor completely subjected to the support.

11.3.3. Self-reflective mediation

This last type of mediation takes heritage as a basis for the projection mapping in order to enhance it. Already, in 2012, the European project *International Augmented Med* (I AM) aimed at enhancing heritage sites through augmented reality, multimedia and interactive techniques. In this context, a projection mapping work was created by Kònic Thtr on the Lebanese citadel of Jbail-Byblos, known as one of the oldest continuously inhabited cities in the world and included on UNESCO's World Heritage List. Projection mapping is a technique used in general event such as festivals, but more often than not it is a matter of place-specific interventions, with the aim of clearly and rigorously transmitting its history. In the case of museums, it allows visitors to access exposed content, such as micro-mapping works.

11.3.3.1. Reconstruction tool

Projection mapping is a perfect alternative for digitally restoring lost colors of a heritage, or restoring its evolution, as in the case of the Romanesque church of Sant Climent de Taüll, in Catalonia. Now a reference, this church has succeeded, thanks to the *Taüll 1123* projection mapping (Figure 11.3) created by Burzon*Comenge and Playmodes, on the one hand to interpret its meaning with a projection that virtually recovers the frescoes that were transferred and exhibited at the Catalan Museum of Fine Arts (MNAC) while exposing the restoration process of the ones that were left.

On the other hand, projection mapping has helped to create a focal point for the rest of the surrounding Romanesque churches that were included in the UNESCO World Heritage List in 2000: "In quantitative terms, the digitalization is a clear success, and it has brought the valley a small taste of a Guggenheim effect of sorts" (Ramos 2018, p. 21). The reconstruction is the result of a scientific process and offers itself as a valid strategy for its rigorous mediation: "By joining together a ruin and a mask, the digital mapping in Taüll is more honest than the previous painted copy of the fresco" (Ramos 2018, p. 17).

Figure 11.3. *Frescos of the apse of the church of Sant Climent de Taüll, in Catalonia, reproduced in the Taüll mapping 1123, 2013 (source: Playmodes Studio + Burzon*Comenge)*

Earlier, Skertzò had set a precedent by virtually restoring the lost colors on Amiens Cathedral from 1999 to 2016. The laser technique helped to recontextualize the church and its sculptures as they were in the Middle Ages. Art historian Santos Mateos (2014) confirms that "the technique of projection mapping is interesting for reconstructing lost or relocated artistic heritage, as long as it combines rigor with spectacularity".

In a museum context, projection mapping has also contributed to virtually restoring the original pigments from the spouses' Sarcophagus to the Bologna History Museum (Fischnaller *et al.* 2015), surrounded by a pyramid made of holograms. By adding storytelling to the exhibition, they created a more emotional connection with visitors. At the Metropolitan Museum in New York, the same year, the Egyptian Temple of Dendur regained its colors thanks to the MET lab team. In 2008, the Ara Pacis, the altar of peace presented in the museum designed by Richard Meier in Rome, also virtually regained its original colours, thanks in part to *The colors of the Ara Pacis* light show. The Katatexilux studio benefited from the collaboration of experts, including those who had contributed to the monument's museographic project, to choose the colors not in a *com'era* state, but based on laboratory analyses and chromatic research on other Roman monuments and paintings.

11.3.3.2. *Museographic tool*

Museums can thus benefit from a tool that does not damage the original pieces and can considerably enrich visitors' understanding of them, either through virtual restitution and reconstruction, or as a complement to other mediation tools in a museum or educational project.

In 2015, Sila Sveta planned a permanent installation in the dome of the Moscow Museum of the Great Patriotic War to tell the story of the process that led to the Great Victory of 1945. It was created as part of the commemorations of the 70th anniversary of that event and covered its most significant episodes. The Jerusalem History Museum at the Tower of David also used projection mapping to celebrate the city's 4,000th anniversary in Skerzò's lighting show *Or Shalem*. The images, projected on the walls, archaeological ruins and paths of the citadel, are inspired by the history of mythical Jerusalem.

Less monumental, micro-mapping works punctuate the exhibitions, as in "Pasteur, the Experimenter" at the Palais de la découverte in 2018, where the agency Les Chevreaux Suprématistes created *Champs maudits* in addition to model projections. This animation, made of wool, is projected on a life-size sculpture of a sheep, to tell the story of Louis Pasteur's investigation into what is now called anthrax. Micro-mappings can also support an educational activity, such as at the Alsatian Museum, as part of the exhibition "Life in mini. Doll Houses & Co." in 2016, in collaboration with Shadok. Based on the tale *Le Fantôme de la Stub*, the 8–14-year-old participants developed a content projected in micro-mapping on a house model:

> "Video mapping as a projected augmented reality can, therefore, have a promising future in education given its potential for the presentation of information in ways that increase the possibilities for conceptual assimilation by

facilitating inference processes and contributing to the transformation of the object (scale model) into knowledge." (Barber *et al.* 2017, p. 338)

Thus, projection mapping allows interactivity with visitors, promoting a more dynamic relationship with the objects or speech on display. "This offers the potential not only to open up heritage to new groups, but also to enable a restructuring of authority and the possibility for a more democratic engagement with history" (King *et al.* 2016). In 2015, in the exhibition "The First Emperor – China's Terracotta Army", the Moesgaard Museum collaborated with the CAVI (Centre for Advanced Visualization and Interaction) to explain to visitors that the terracotta warriors at the Emperor Qin's tomb were actually colored (Figure 11.4). Visitors were offered a selection of colours similar to those that originally adorned them, so they could choose how they wanted to dress the replica that served as a model by projecting their choice through projection mapping. Once the virtual coloration was completed, the model could join the digital army behind it and let the next visitor continue the ongoing process.

Figure 11.4. *Terracotta warriors rediscover their colors in the exhibition "The First Emperor – China's Terracotta Army". at the Moesgaard Museum, 2015 (source: CAVI, Aarhus University)*

A similar principle was applied during the "Making Mainbocher" exhibition at the Chicago Museum of History in 2016–2017. The interactive installation "Design a Dress" allowed the visitor to virtually dress a mannequin using a touch screen by choosing the patterns and styles of dresses that were applied to it. This time, it is not a question of understanding a historical fact, but of experimenting freely on the theme of the exhibition, which has the effect of enriching or heightening the experience. Other museums, such as the V&A Museum or the Chicago Museum of Science and Industry, have used projection mapping in conjunction with other technologies to allow visitors to experience and understand the world in which we live by imagining our future through these experiments.

Other museums use projection mapping to create an atmosphere, to enhance the experience, as in the case of the *Memory Collector* (Figure 11.5), which Moment Factory Studios created at the Pointe-à-Callière Museum in Montreal. The first collector sewer built in North America was transformed, in this case, into an archival support representing the memory of Montrealers, while wanting to inject poetry into the place in a soothing setting that transports us. It is also a question of offering a visitor experience designed to be different by using immersive techniques in the tunnel that leads to the projection. As a technological tool, projection mapping has the potential to have a "cultural value of digital involvement" with heritage (King *et al.* 2016).

Figure 11.5. *Memory Collector highlights the structure of the first collector sewer built in North America as well as the memory of Montrealers, while providing an experience for visitors to the Pointe-à-Callière Museum in Montreal (source: Moment Factory)*

The use of projection mapping also follows the proliferation of "experiential" type exhibitions, the immersive exhibitions that are not based on collections, of which Vincent Van Gogh and Klimt represent the main protagonists. The Atelier des Lumières is one of the pioneers in marketing the concept (which has provoked several museological debates), where sensations prevail over the experience of the original, and where the visitor is totally enveloped by different aspects of the digital works of the artists on display. A fluid animation is projected on all the architectural elements of the space, walls, ceiling, floor and even columns where the works are intertwined with different motifs and movements resulting from them, all accompanied by an original musical creation.

These proposals are of uneven quality, depending on the degree of immersion, the staging and the way the works are shown. Because it does not replace the original collections, this strategy cannot be compared to a classical exhibition with the same objectives, but it can be argued that it can be seen as a complement to the exhibition by offering an introduction to visitors for whom the museum seems inaccessible.

In the context of an exhibition, projection mapping therefore also offers several kinds of mediation, ranging from history to didactics, sensory and entertainment.

11.4. Conclusion: monumentalize the monumental

Video and micro-mapping can be projected almost anywhere, and can interact with other heritage contexts, such as gastronomic heritage or living heritage (such as theatre). In this text, only interactions with built heritage and museums have been analyzed in order to identify three main types of mediation: the first mediation, in terms of influence, involves content external to the place affixed to it in order to disseminate it, taking advantage of the iconic status of the heritage, or its position value. The danger with this type of interaction is to use the heritage for promotional purposes only, and to detach the message completely from the medium. Heritage, in this case, is instrumentalized, and this is what can lead to rejection or distort the spirit of place. The second mediation identified is combined mediation, which involves a fusion between the artistic act of projection mapping and the heritage place on which it is projected. The projection highlights it by evoking some of its aspects, formal or symbolic, without however being part of a frank approach of mediation. Finally, the third mediation, which is self-reflective, uses video mapping as a cultural mediation device, in the sense that it allows visitors to a heritage site to transmit its history or to understand its mechanisms in an interactive approach, which can go as far as the dematerialization of collections.

This outline of a mediation typology gives an overview of different interventions between projection mappings and heritage. The types of mediation were analyzed according to the relationship that the narrative (or abstract) content has with its heritage medium. In continuity with this study, an in-depth analysis of the formal means as well as the sound and music used would shed more light on this relationship.

The cases chosen to illustrate the types of mediation have a common characteristic: most video mappings are projected on monumental heritage, in the sense that it has acquired this status in an established consensus; they are often grandiose, impressive or of undeniable importance. Consequently, there is a double monumentalization: first, projection mapping monumentalizes in the architectural sense, by "putting [heritage] in self-representation" (Debray 1999, p. 35). But also, it monumentalizes in a cultural sense, by "injecting meaning". When projection mapping is projected in an urban context, it can serve as a "marking" point (Veschambre 2007) in order to reclaim a social space. In other words, projection mapping can reactivate heritage, or a memory trace, to update it through creative action, thus affirming spatial appropriation.

Projection mapping works could also be a great tool in areas that require enhancement, rehabilitation, and not only to support their iconic status. Quebec artist Danny Perreault, for example, used this technique to reinvest an abandoned fountain in Senegal, a relic from the colonial era, "once erected as a symbol of freedom and emancipation" (Perreault 2017). In this way, he follows the tradition of Krzysztof Wodiczko, whose works have a social and not only spectacular impact.

Projection mapping offers possibilities of mediation because it "appears as a spectacle in a post-identity context, where the processes of creating images in real-time such as live coding or projection mapping are presented as a memory technique (or recollection), at the same time as a vertiginous spectacle of optical illusions" (Alonso 2013). The challenge is to find this balance between the two.

Today, some forms of data art constitutes a continuity of projection mapping when it blends with architecture, as was the case in September 2018 on the Walt Disney Concert Hall. The artist Refik Anadol projected almost 45 terabytes of digital data from the orchestra's digital archives onto the iconic building. In this case, though the technique is essentially similar, the medium takes from digital data. Therefore, it deepens the interaction between the form of the projection, the location and meanings of heritage, as well as its readability. The innovations of projection mapping would thus be found in the monumentalization of small heritage or in micro-mapping, in order to bring a specific content complementary to other mediation tools.

11.5. References

Alonso Atienza, L., Gárciga Romay, L. (2013). "Qué gigantes" dijo Sancho Panza. Proyecciones monumentales con Video Mapping en los bicentenarios de las independencias de las naciones latinoamericanas. *Arte y sociedad*, 4.

Barber, G., Marcos, L., Accuosto, P., García Amen, F. (2017). Interactive Projection Mapping in Heritage: The Anglo Case. In *17th International Conference, CAAD Futures 2017*. Istanbul, Turkey, 337–348.

Choay, F. (1999). *L'allégorie du patrimoine*. Le Seuil, Paris.

Davallon, J. (2004). La médiation: la communication en procès? *MEI*, 19, 37–59.

Debray, R. (1999). Trace, forme ou message? *Les Cahiers de médialogie*, 7, 27–44.

Fabre, D. (ed.) (2013). *Émotions patrimoniales*. Fondation Maison des Sciences de l'Homme, Paris.

Fischnaller, F., Guidazzoli, A., Imboden, S., et al. (2015). Sarcophagus of the Spouses installation intersection across archaeology, 3D video mapping, holographic techniques combined with immersive narrative environments and scenography. *Digital Heritage*, 1, 365–368.

Georgescu Paquin, A. (2014). *Actualiser le patrimoine par l'architecture contemporaine*. Presses de l'Université du Québec, Quebec.

Georgescu Paquin, A. (2015). Les mappings vidéo sur le patrimoine bâti comme forme d'hybridation entre œuvre et dispositif de médiation. *Études de communication*, 45(1), 53–75.

Hanson, K.M. (2014). Video Mapping: A Medium for Projection Artists. In *Encyclopaedia Britannica 2014. Book of the Year*. Encyclopaedia Britannica, Chicago, 198–199.

Hetsroni, A., Tukachinsky, R.H. (2005). The Use of Fine Art in Advertising: A Survey of Creatives and Content Analysis of Advertisements. *Journal of Current Issues and Research in Advertising*, 27(1), 93–107.

King, L., Stark, J.F., Cooke, P. (2016). Experiencing the Digital World: The Cultural Value of Digital Engagement with Heritage. *Heritage & Society*, 9(1), 76–101.

Krautsack, D. (2011). 3D Projection Mapping and its Impact on Media & Architecture in Contemporary and Future Urban Spaces. *Media-N, Journal of the New Media Caucus*, 7(1).

Kronhagel, C. (2010). *Mediatecture: the Design of Medially Augmented Spaces*. Springer, Berlin.

Mateos Rusillo, S.M., Gifreu-Castells, A. (2014). Reconstrucción y Activación del patrimonio artístico con tecnología audiovisual. Experiencia de Taüll 1123. *El Profesional de La Información*, 23(5), 527–533.

Mateos Rusillo, S.M. (2019). *Miradas desde la copa, e-Revista de Comunicación y Patrimonio cultural* [Online]. Available at: www.comunicacion patrimonio.net/.

Morisset, L.K. (2009). *Des régimes d'authenticité: Essai sur la mémoire patrimoniale*. Presses de l'Université du Québec, Quebec.

Perreault, D. (2017). Le vidéomapping, un art du trompe-l'œil et de l'anamorphose au potentiel social et communautaire indéniable. *ETC MEDIA*, 111, 36–39.

Ramos, C. (2018). Aura and Spectacle: The Digital Restitution Project at Sant Climent de Taüll. *The International Journal of New Media, Technology and the Arts*, 13(2), 11–23.

Raskar, R., Welch, G., Cutts, M., Lake, A., Stesin, L. (1998). The office of the future: a unified approach to image-based modeling and spatially immersive displays. In *25th annual conference on computer graphics and interactive techniques. Siggraph '98*. ACM, New York, 179–188.

Salas-Acosta, L.-M. (2014). Arquitectura Artificial Y Sostenibilidad: Proyección Vídeo Monumental. Universidad de Murcia, Murcia.

Veschambre, V. (2007). Le processus de patrimonialisation: revalorisation, appropriation et marquage de l'espace. *Café géo* [Online]. Available at: http://cafe-geo.net/wp-content/uploads/processus-patrimonialisation.pdf.

12

Projection Mapping: A Mediation Tool for Heritage Resilience?

12.1. Introduction

Architectural works are the most telling traces of human civilization. Whatever the hands that have traced them, they are the bearers of materialized, visible and meaningful cultures. Practical, social and aesthetic functions as well as architectural signs are an integral part of World Heritage, marked not only by technological progress, but also by a structuring of the significant units of architecture, which gives works a cultural dimension, social significance and a spirit of aesthetic renewal (Zhang 2009).

Nevertheless, according to the architect Marika Jacquemart-Bouaoudia[1], who is also a guide interpreter, the public prefers to be interested in painting, literature, sculpture or music, rather than architecture. One reason is that these areas are more accessible and the tools to understand them are easier to assign. In other words, it is easier to hold a visual art exhibition, a book fair or a concert than to hold an architectural exhibition because the objects are not very movable. To be interested in architecture, you have to go towards the objects through guided tours and commented by knowledgeable guides (expert guides)!

Chapter written by Hafida BOULEKBACHE and Douniazed CHIBANE.
1 http://ch.viadeo.com/fr/profile/bouaoudia-jacquemart.marika.

Because of this lack of interest in architecture, Bruno Zévi[2] has written a book entitled *Apprendre à voir l'architecture*, in which he recommends educating the eye, to exercise the eye on volumes, shapes, solids and voids. For each architecture, it is necessary to identify and decipher the internal and external spaces and make each person an informed but positive critic. Bruno Zévi also insists on the importance of considering the times and context of the buildings.

He also points out the sensitive values of architecture; the value of an architectural work does not differ according to the period of its construction. In other words, it should not be argued that a building is beautiful because it is old, or on the contrary that other building is ugly because it is modern. To evaluate the beauty of a building, it is necessary to be able to see the volumes, proportions and harmony of the facades, but also to feel the internal space that subjugates, attracts and elevates against those who push back and tire.

12.2. Architecture, a heritage trace and an art to be preserved

Architecture is often devalued because it is poorly understood, denigrated because it is poorly viewed. However, many theorists and architects consider it as an art in its own right, since, in addition to being conditioned by technology, it aspires to an aesthetic ideal.

According to Jay A. Pritzker[3] and his wife, founders of the Architecture Prize of the same name, "architecture is intended to transcend the simple need for shelter and security by becoming an artistic expression". Jonathan Jones[4], a renowned art reporter for *The Guardian*, who was also a jury member for the 2009 Turner Prize, says that "architecture is the art we encounter most often, most intimately, but because it is functional and necessary for life, it is difficult to be clear, where in a building, where art begins".

---

2 Bruno Zévi (Rome, 1918-2000), architect, historian and art critic. After graduating from Harvard (USA) in 1941 under the direction of Walter Gropius, he returned to Italy and, in 1948, taught history at the University Institute of Architecture in Venice. The same year, he published *Saper Vedere l'architettura*, which became a best-seller translated into 15 languages!
3 https://archibat.com/blog/3179/.
4 https://archibat.com/blog/3179/.

When you look at the Singapore Museum of Scientific Arts, this building that looks like a huge white lily that comes out of Marina Bay (Figure 12.1), and whose ten huge petals can capture rain and release it in one place, you can only agree with Le Corbusier who explains: "We use stone, wood, cement, we make houses out of it, palaces are construction. Ingenuity works. But, suddenly, you take me to my heart, you do me good, I'm happy, I say: it's beautiful. That's the architecture. Art is here.[5]"

Figure 12.1. *Museum of Scientific Arts, Singapore (source: https://archibat.com/blog/3179/). For a color version of the figures in this book see, www.iste.co.uk/schmitt/image.zip*

According to Le Corbusier, architecture is the result of a "skillful, correct and magnificent interplay of volumes assembled under light. Our eyes are made to see shapes under light; shadows and lights reveal shapes; cubes, cones, spheres, cylinders or pyramids are the large primary forms that light reveals well; the image is clear and tangible to us, without ambiguity. That's why they are beautiful forms, the most beautiful forms[6]". These volumes, whatever the period of their construction, these vestiges of the past or cathedrals of modern times, are testimonies and carry within them a communicative trace, an intentional expression; a heritage value.

Is this trace visible to everyone at all times? The architectural heritage, this route inherited from the past or erected in our daily lives, from which we

---

[5] Le Corbusier, 1923, Vers une architecture, https://web.ac-reims.fr/dsden10/exper/IMG/pdf/rencontre_lecorbusier.pdf.
[6] https://dicocitations.lemonde.fr/citations-auteur-le_corbusier-0.php.

benefit today and which we pass on to future generations, is becoming increasingly vulnerable through threats linked to natural disasters such as earthquakes, floods or social disasters such as wars and revolutions. Its importance as a resource for cultural and economic development is at the heart of current debates, all of which converge on the need to safeguard and transmit it, but also on the urgency of its mediation (Figure 12.3). This mediation is taken here in its broadest sense. In other words, it refers to any action that promotes the encounter between the work of art and its recipient (Caillet *et al.* 2000).

Figure 12.2. *Vuitton Foundation*

Figure 12.3. *Remains of Timgad (source: Araibia Amine)*[7]

---

7 Timgad or Thamugadi (Marciana Traiana Thamugadi colony in Latin), nicknamed the "Pompeii of North Africa", is an ancient city located on the territory of the eponymous municipality of Timgad, in the wilaya of Batna in the Aurès region, in northeast Algeria.

12.3. The architectural heritage between preservation and mediation issues

Beyond the challenge of conserving architectural heritage, the necessary transmission to future generations also implies a reflection on the mediation process to be put in place. In this sense, digital mediation devices are multiplying: from the touch table to *in situ* augmented reality, websites and mobile applications, wide access to heritage is now possible.

Digitization therefore seems to meet this safeguarding challenge and perpetuates the trace of these past resources. The example of Syria, where several digitization operations have been carried out on archaeological sites, can attest to this[8].

In the context of architectural heritage, digital technology and new forms of induced writing produce new forms of representation and mediation. The use of new tools and media has resulted in the evolution of uses that, in fact, were born thanks to digital technology. However, several conditions must be met in order to hope for mediation representation:

– a materiality: need for an eminently material support to be able to exist;

– intentionality: a purpose that gives meaning to what one wants to represent and a means of welcoming it;

– a temporality: linked to a more or less long process;

– an interpretation: since the device challenges, calls for curiosity and arouses interest;

– reflexivity: this refers to reactivity, feedback, emotion, experience.

12.4. Meeting between architectural heritage and projection mapping

Projection mapping[9], also known as "illusionist projection", is a projection technique that consists of projecting a video adapted to the size of

---

8 Syrian heritage database, the result of a vast digitisation operation by the French start-up Iconem, with the Direction générale des antiquités et des musées syrienne (DGAM).
9 The term "projection illusionist" was proposed by the Office québécois de la langue française in 2012 to designate this concept: http://gdt.oqlf.gouv.qc.ca/ficheOqlf.aspx?Id_Fiche=26519865.

the medium that receives it, a medium whose primary use is not that of the screen.

Etymologically, projection mapping is composed of the term *mapping* meaning to map which joins the term video to refer to this technique. At the crossroads of photography, music, graphics, coding and engineering, this new discipline of projection mapping is as much a part of digital art[10] as it is of mediation and has developed since the 1990s, the years in which digital arts entered the art world.

To produce the mapped video work, a mastery of digital tools is necessary, but other prerequisites such as architectural knowledge are essential. Projection mapping requires a good knowledge of the medium; that is the nature of the receiving structure, its dimensions, angles, shapes, etc., since all these elements must be taken into account to model a 3D model on a computer, and then allow us to imagine the desired projections (Rautureau 2014).

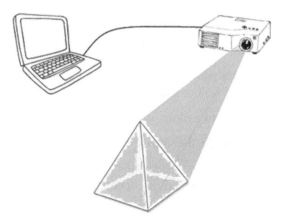

Figure 12.4. *Projection mapping (source: A. Pogoda)*

Often used in festive or commemorative events, this technique is also used in other contexts. Whether in confined spaces, inside museums or in an urban environment, it is becoming exponentially more democratic today. In addition to being considered a show, projection mapping is also a useful tool for a mediation process of architectural heritage.

---

10 Digital art refers to the artistic activity that uses digital language as the material of its language: https://perezartsplastiques.com/2015/10/15/lart-numerique/.

The conceptions around these projections can be very varied and more or less rich, as any result of a creative process. Nevertheless, these projections could be briefly classified according to the expected objectives and target to be achieved.

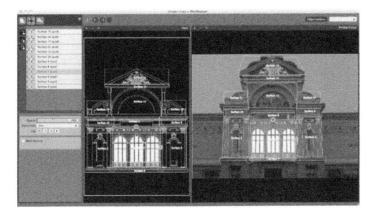

Figure 12.5. *Madmapper, projection mapping software[11]*

12.5. Classification of architectural projection mapping

12.5.1. *Communication issue*

12.5.1.1. *Projection mapping for events*

Projection mapping is a technique that is very popular with municipalities that plan to communicate about an event or a commemoration. Each large city has its own "projection mapping meeting" and several examples can be mentioned. Below are some illustrative examples (Figure 12.6).

Figure 12.6. *Celebration of lights in Lyon[12]*

---

11 Source: www.legatomusiccenter.com/store/madmapper.
12 Source: ww.lonelyplanet.fr/sites/lonelyplanet/files/media/article/image/hotel-ville-lyon_0.jpg.

12.5.1.2. Projection mapping for marketing

Projection mapping is very popular in marketing because it offers unparalleled visibility support for products. Advertisers have understood the communication challenge of this technique and some brands, particularly car manufacturers, are very fond of projection mapping (Figure 12.7).

Figure 12.7. *The Lexus advertising campaign*

12.5.1.3. Interactive projection mapping

Some projections integrate the viewer and stage him in the sense that he becomes an actor in the process. Perspective lyric with reference to the lyrical arts that held a prominent place in the "Grand Théâtre" is an interactive projection mapping project carried out by the *1024 Architecture* label on the façade of the Théâtre des Célestins for the Festival of Lights in Lyon in 2010 (Figure 12.8). Based on the voice and song of the audience, a system of vocal visual distortions allows the viewers to intervene on the projection according to the words, songs, noises they emit, and thus give the face a specific shape indicating the existence of a dialogue between the work and its audience. This project was very successful and was therefore exported to Neuchâtel in Switzerland and Singapore[13].

---

13 www.1024architecture.net/projects/?filter-state=France.

 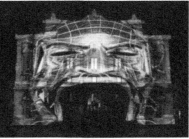

Figure 12.8. *Lyrical perspective in Lyon (source: 1024 Architecture)*

12.5.1.4. Commemorative projection mapping

Another example is the city of Prague, which used this technique in 2010 to celebrate the 600th anniversary of the astronomical clock located in the Old Town Square in the center of Prague. As the monument is known to be a historic monument much visited by tourists, projection mapping is used here to commemorate and tell the trajectory of the city through several sequences, temporal narratives and historical moments, and symbols, linked to these different stops in the past, are projected (Figure 12.9).

Figure 12.9. *Projection mapping on the astronomical clock in Prague (source: 1024 Architecture)*

12.5.2. *Information issue*

Projection mapping brings new paradigms to the fore: urban-scale support and immersive experience. This fertile crossover promotes access to more and more data, often unintentional. Yet, these "inscriptions" produce information; information that can be spread with a framework message that would be "intentionalized" by metacommunication. Provoked according to the designer's design, the information is structured using the technical device and its component artefacts are placed in maximum visibility while others are placed in the background, in order to produce the desired meaning. The objective is then to promote the processes of identification and cognition, the digital device is considered as a tool for adding information and cognitive extension.

12.5.2.1. *Projection mapping in Seville: work by painter Murillo*

In this example, the objective was to make the works of the Sevillian painter known through a didactic projection mapping using the casino facade. These works were highlighted for 10 days with three screening sessions per day (Figure 12.10).

Figure 12.10. *Mapping on the works of the painter Murillo in Seville*

12.5.2.2. *Mapping on the Seville City Hall*

This time, another didactic projection provided a fascinating opportunity to explore the history, art, culture and tradition of Christmas and its symbols.

This allowed more than 500,000 people to discover these informative elements on the façade of Seville's city hall (Figure 12.11).

Figure 12.11. *Projection mapping of Seville City Hall*

12.6. Meeting between architecture and projection mapping

By paraphrasing Paul Ricoeur, we can argue that architecture is a document to be preserved. It is to be preserved because it includes both "written" witness, in the visible order, and "unwritten", in the sensitive order.

Architecture stands out from other heritage objects by its scale and its ability to blend into public space. It exists in a 3D space that includes man. However, architecture is often poorly understood, misunderstood and therefore carelessly preserved.

To see it better, to learn to look up, to discover the facades, corbels, etc., and other components of architecture, projection mapping is becoming the learning device, the engine of change. In addition to being a stimulating and recreational tool, it offers a new setting for buildings and architectural objects that are part of our daily lives and to which we do not necessarily pay attention. Projection mapping thus reinvents architecture since it is used as a support, but at the same time serves as a creative material.

This marriage of reason between projection mapping and architecture brings out the significant dimension of these objects. In addition to putting an imprint on a façade, the scenic design of this projection can build intelligible from the sensitive and generate a movement of thought that penetrates the building and its components. This building, which helps to immerse the audience in another space-time, accentuates reflexivity and memorization.

Indeed, this digital device makes it possible to point out the architectural elements (measurements, distances, extent), the combination of all the parameters (shape, material, color, etc.), the arrangement of the constituent parts (area, partition, structure, wall, piercing, roof, etc.), the coherence and unity of the parts and all the subsystems and the balanced link with the "architecture" system. The interplay of relationships superimposed on the lights, the sequences of relationships, constitute the thought of architecture and are the determinants of its conception and the constructs of its meaning. Beyond the classical objectives (cognitive, experiential and institutional), projection mapping is a sign of a reflexive content on the scenic exposure of architecture.

12.7. Conclusion

> "Architecture is not only an art, not only the image of the hours spent, lived by us and others: it is first and foremost the setting, the scene where our life takes place."[14]

The architecture is supported by a material object, visible and endowed with a particular potential for meaning. This object is endowed with a singular materiality which, while pointing out the present, evokes an absent but postulated past. It is raised and interpreted because it integrates a purpose. But with projection mapping, this mediation tool, architecture as well as being a projection medium becomes part of the work. It is a representation and a support of the enunciation. This crossing generates phenomena of perception and judgment by evoking the "unspeakable" of sensoriality.

Projection mapping "heightens" the architectural reality; while respecting it by adapting to the building, it magnifies it and makes it possible to highlight traces that have disappeared, through iconography, sound, light and narrative that come to settle on the surfaces. Through its immersive character, it offers the viewer a powerful theatrical staging where the façade

---

14 www.academia.edu/34566801/Architecture_quotes.

is no longer in 2D, but multiplies in several dimensions (3D, sound and light) that penetrate the architecture and make it discover in all its thicknesses until reaching the framework.

This technique respects the architecture and is forgotten by hiding the whole technological arsenal, to make the stones speak and insist on their historical importance.

Projection mapping offers architecture the possibility to rethink the processes of externalization of architecture. Through these new artefacts, architecture manages to be better seen, to fight against its erasure and becomes a theatre of memory, thus a heritage resilience. Projection mapping is a mediation tool that perfectly resonates with the draft operational guidelines adopted by the Committee of the UNESCO Convention for the Protection and Promotion of Cultural Diversity in 2005, relating to the impact of digital technology on the diversity of cultural expressions, since it materializes, makes it possible to study, remember, commemorate and represent it.

Projection mapping offers real possibilities of heritage mediation, especially when projected on symbolic buildings, (Georgescu Paquin 2015). Indeed, the inclusion of projection mapping in the architectural heritage allows for a mediation that is certainly short and of a reduced duration, but which can arouse the interest of the public to come (re)discover the heritage and to be made aware of the induced meaning. It is therefore important to take a closer look at the projected content. In this case, the scenario of the projection mapping project must not only aim at the aesthetic aspect, but also be a vehicle for an awareness message.

12.8. References

Bordeaux, M.-C., Caillet, É. (2014). La médiation culturelle: pratiques et enjeux théoriques. *Culture et musées*, special edition, 139–164.

Caillet, É., Pradin, F., Roch, E. (2000). *Médiateurs pour l'art contemporain : répertoire des compétences*. La Documentation française, Paris.

Diouf, L., Vincent, A., Worms, A. (2013). Les arts numériques. *Dossiers du CRISP*, 81(1), 9–84 [Online]. Available at: www.cairn.info/revue-dossiers-du-crisp-2013-1-page-9.htm.

Fraysse, P. (2015). La médiation numérique du patrimoine: quels savoirs au musée? *Distances et médiations des savoirs*, 12.

Georgescu Paquin, A. (2015). Les mappings vidéo sur le patrimoine bâti comme forme d'hybridation entre œuvre et dispositif de médiation. *Études de communication*, 45, 53–76.

Le Corbusier (1923). *Vers une architecture*. Flammarion, Paris.

Rautureau, M. (2014). Vidéomapping: pratiques contemporaines d'un nouvel art élargi en France. Thesis, Université Sorbonne Nouvelle, Paris.

Welger-Barboza, C. (2001). *Le Patrimoine à l'ère du document numérique. Du musée virtuel au musée médiathèque*. L'Harmattan, Paris.

Zévi, B. (1959). *Apprendre à voir l'architecture*. Éditions de Minuit, Paris.

Zhang, X. (2009). Approche sémiologique de l'architecture. Synergies Chine, 4, 205–214.

13

Architectural Projection Mapping Contests: An Opportunity for Experimentation and Discovery

13.1. Introduction

There are several ways to organize architectural video projections. Apart from commercial proposals and municipal orders for festivities, there are international projection mapping contests.

These contests in the form of calls for projects are real platforms for meetings and development of the projection mapping sector. Indeed, they confront different styles through original proposals from competitors, who come from all over the world and who impregnate their creation with their respective cultures. They thus provide opportunities for artistic experimentation and risk-taking where everyone stands out for their expressiveness and hopes to win the prizes.

With these festivals, each year more and more numerous, artists only have the creative aspect to develop. Indeed, the organizer takes care of the technical installation, distribution and preparation of the files necessary to create the projection mapping on the building, including the 3D model and reference photos. These international festivals feature a network of calls throughout the year. The oldest, which still exist, are the Circle of Light in Russia (est. 2011), the 1-Minute Projection Mapping in Japan and the Genius

---

Chapter written by Jérémy OURY.

Loci Weimar in Germany (est. 2012). Their organizers are now claiming to receive more than 150 proposals each from more than 30 nations.

These festivals offer a real artistic variety because the projects, sometimes submitted to a theme, use various aesthetics and creative techniques. The public can then, like a museum, confront several styles and narratives on the same facade.

Participating in these projection mapping contests since 2015 as a competing artist, I have been able to establish contacts with both organizers and artists. As I am constantly monitoring the evolution of projection mapping projections on buildings on an international scale, I present these new places of emergence and experimentation through a participant observation of these contests which gather about 20 regular competitors. This reflection is also based on three interviews with the organizers Tatyana Popova, Hendrik Wendler and Michiyuki Ishita of the Circle of Lights, Genius Loci Weimar and 1 Minute-Projection Mapping festivals.

13.2. Different projection mapping projection contexts

13.2.1. *Limitation of projection mapping orders*

There are different frameworks for making architectural projection mappings. There are several private possibilities such as commercial projection mapping works set up for product launches, brand promotion or during a particular event. On the side of public organizers, there are activities organized by elected officials to highlight a city's heritage, commemorate festivities, such as winter festivities, or occasionally make a city attractive with one or more projections on prestigious buildings. In both cases, there is a notion of a client/commissioner relationship with the artist who must provide an artistic or technical/artistic performance according to precise specifications. In France, these screenings are organized as public markets or as direct commissions to artists. This implies a certain level of legal and economic structuring and a certain recognition in the community, particularly in the case of an order, to access such projects.

However, these contexts are not conducive to the emergence of new artists or new practices because financial stakes reduce risk-taking. In addition, the complexity of the specifications, which include both technical and artistic aspects, means that service companies are in the best position to respond to these calls. They already work with their own creative design office or with well-known artists such as Damien Fontaine, Les Spectaculaires – Allumeurs d'Images, Skertzo or Ad Lib Creations who have carried out a large part of the projection mappings in French cities since 1999. Most of these projection mappings have the constraint of having to be a "family" show, that is accessible to everyone and especially to children. A narrative aspect involving the history of the city or building is often requested through a voiceover. The historical framework can also be developed with animations of characters on the facade accompanied by epic music of symphony orchestra or electronic music. These projection mappings are generally of a standardized duration of 10 to 20 minutes, broadcast in a loop over a period of one to several weeks to be seen by the largest number of residents and tourists. These specifications relegate the notion of projection mapping as a work of visual and sound art to the background in favor of the spectacular.

The style is struggling to be renewed because more and more cities are offering these "sound and light" shows based on the same concept: projection of a historical projection mapping on the cathedral or the town hall. For example, there are only a few festivals dedicated to artistic projection mapping in France compared to a high number of "sound and light" events.

13.2.2. *Contests, platforms of creative freedom*

The gradual development of the structures of the digital arts sector has enabled, among other initiatives, the emergence of open call for projects and contests from international festivals. They provide real opportunities for the artistic development of projection mapping. Indeed, the organizer is launching an open call for participation in the form of an artistic contest asking several artists to prepare a visual creation for the festival. He is in charge of the technical side of the projection since the festival takes care of the rental, installation of equipment and media distribution. He also provides artists with reference photographs, including blueprints and masks specifying the parts of the building on which the projection will be made, and sometimes produces a 3D model of the building. Both amateurs and professionals can participate in the contests and only have the creative aspect to develop. This simplifies the time spent working on the project and places the competitor in the position of an artist without taking any financial risk for him.

These festivals, which include these contests, are now the main space for the expression and exhibition of projection mapping works. Many festivals dedicated to the digital arts fully play this role as sponsors of works, producers and presenters. Indeed, these structures, true platforms for presenting the work of artists, encourage them and allow some of them to emerge. These festivals offer a real artistic variety because they select several screenings using different creative techniques and cultural influences from several continents. These contests thus become meeting places for international artists from diverse backgrounds.

These projects are also beneficial to the spectators since during the same evening, the public can compare different visual and sound artistic proposals. By inviting spectators to vote and decide between artists, these festivals also educate the public, which is becoming increasingly demanding.

13.3. Interests and functioning of the contests

13.3.1. *The organizers' point of view*

The 1 Minute Projection Mapping is a unique contest with a specific format, since it requires a creation of a maximum of one minute. Since 2012, it has been proposing a different building as a projection surface so that artists can participate in each edition. As explained by Michiyuki Ishita, the festival director, the objective is to give artists from all over the world a creative space to try out new graphic proposals on a building thanks to the technical support provided by the festival. Indeed, it is not easy to have access to large architectures to do architectural projection mapping when you are an artist. It is important to test your creations because the results of the projection can be very different from the initial proposal. Finally, projects are created on a computer screen while they are intended to be projected onto raised surfaces. The festival therefore gives artists the opportunity to visualize these works on a given architecture. Finally, commentary from the public and a jury of professionals on artistic creation is a valuable feedback for the artist.

For Tatyana Popova, from the organizing team of the triple contest Art Vision: classical, modern and VJing, integrated into the Circle of Light festival of Russia, the contest gives artists the opportunity to make a name for themselves based on their artistic know-how. She also explains that it is a question of showing the public the art of projection mapping without influencing the proposals. Going through a contest rather than a commission removes the budgetary limitation and does not impose a predefined artistic style.

Since 2012, with the Genius Loci Weimar, Hendrik Wendler has been focusing both on commissions that guarantee style and quality for his festival, and on a call for projects that introduces a certain amount of surprise and unexpected surprises into the program. Although this represents a risk, the organizer gives a chance to new artists and new forms by opening opportunities to a very fertile sector. Nevertheless, the festival attaches great importance to the theme that anchors its event in the historical, political and cultural context of the city of Weimar.

The artistic diversity of the proposals makes each contest a real ephemeral open-air museum. For example, at the Girona International Mapping Festival, FIMG, in 2015, there were about 30 projection mappings projected on four buildings in the city, representing more than two hours of artistic proposals. These festivals exist throughout the world and establish a network of calls for projects that are spread over the year. Unfortunately, some festivals are not sustainable and new ones are created every year. Some of these include the Circle of Light in Russia (since 2011), 1-minute Projection Mapping in Japan (since 2012), Genius Loci Weimar in Germany (since 2012), Video Mapping Contest in France (since 2014), IMAPP in Romania (since 2014), FIMG (2014–2016) and Luz y Vanguardias in Spain (since 2016) or Zsolnay Light Art in Hungary (since 2016).

13.3.2. *Functioning of the contests*

Each contest has its own specific features and conditions evolve over the years. Several festivals have established a pre-selection based on three concept images from the beginning of the contest to avoid the artist working at a loss on a visual creation. Genius Loci Weimar receives 30-second visual proposals that outline the artistic project. The organizers then make a selection of the projection mappings that will be screened during the event. A jury will then award marks according to different criteria such as: originality, graphic quality, play with the building's architecture, appropriation of a possible theme, the narrative dimension of the project or the relationship between sound and visuals in order to award one or more prizes.

The winning works are often global works with an original style or a singular approach that relegates the classic successions of 3D effects to spectacular. The search for meaning, emotion, coherence, narrative writing and originality is clearly encouraged by the choice of juries. The Circle of Light 2017 Art Vision Classic contest, in a wide variety of styles (Figure 13.1), awarded the work of Ukrainian Julia Shamsheieva with a graphics-cartoon universe, perfectly adapted to the architecture of the Bolshoi Theater. The artist Fluid, winner of the Zsolnay Light Art 2018, proposes a truly poetic work with *Seven* by appropriating the graphic charter of the contest. The Luz Y Vanguardias festival was able to highlight the coherence of the association of sound and video with the projection mapping *Prism* of Studio Echelon Mapping in 2017. While the 2015 edition of FIMG rewarded the writing of the minimalist narrative work of *Nihils Unus* by Filip Roca and *Geren-Tertia*, which combines the world of Anxo Lopez's drawing with that of the Theremin Tourette collective in 2014.

Figure 13.1. *(a) Overview of the various contest projects at the Circle Light Festival 2017. From left to right: Julia Shamsheieva, Global Illumination, Shuka Studio, Romera, VJ Fader & Players, Chema Siscar, Antaless Visual Design, Ouchhh, Skgmedia. For a color version of the figures in this book see, www.iste.co.uk/schmitt/image.zip*

Figure 13.1 (continued). *(b) Overview of the various contest projections at the Circle Light Festival 2017. From left to right: Julia Shamsheieva, Global Illumination, Shuka Studio, Romera, VJ Fader & Players, Chema Siscar, Antaless Visual Design, Ouchhh, Skgmedia*

The remuneration of artists is based on a price scale defined by the functioning of the contest. This can be a ranking of the first three projects, or a jury prize. Often, the public is also invited to vote for their favorite project.

While some festivals pay for artists, most do not have the means to bring them in. However, the presence of the contest participants during the screenings is a sign of real recognition of the work accomplished and provokes artistic and professional encounters. Finally, most festivals broadcast the screenings live and then upload a high-definition recording of the screenings to provide promotional material for each participant.

13.4. Analysis of the 2018 season

13.4.1. *Perspective of the artists*

Since the beginning of 2018, the *List of international video mapping open call & contest 2018* group has gathered more than 850 members on the Facebook platform. The announcements of more than 19 contests and 16 calls for projects around projection mapping projects were relayed (Table 13.1).

Festival	Continent	Country	Type
Rio Mapping Festival	Americas	Brazil	Call for projects
Art on the Mart	Americas	USA	Call for projects
Melody Maker & New Media	Americas	Mexico City	Call for projects
Borealis	Americas	United States	Contest
SSA MAPPING	Americas	Brazil	Contest
SINO NIIO Art Prizes	Asia	Hong Kong	Call for projects
1-minute Projection Mapping festival 2018	Asia	Japan	Contest
1-minute Projection Mapping festival 2019	Asia	Japan	2019 Contest
Okazaki Tokiakari 2018	Asia	Japan	Mini contest
Pomezia Light Festival	Europe	Italy	Call for projects
Amsterdam Light Festival	Europe	Netherlands	Call for projects
Light nights in Gatchina	Europe	Russia	Call for projects
Live Performer Meeting	Europe	Italy	Call for projects
Light Move Festival	Europe	Poland	Call for projects
BAM festival	Europe	Belgium	Call for projects
Limelight Academy	Europe	Hungary	Call for projects
Swiss light festival	Europe	Switzerland	Call for projects
Ibiza light festival	Europe	Spain	Call for projects

Weimar Genius Loci	Europe	Germany	Contest
Novi Light Sensation	Europe	Italy	Contest
Kyiv Light festival	Europe	Ukraine	Contest
Spotlight	Europe	Romania	Contest
Video mapping contest	Europe	France	Contest
La Grua Encantada	Europe	Spain	Contest
Luz Y Vanguardias	Europe	Spain	Contest
M3D festival	Europe	Italy	Contest
Zsolnay Light Art	Europe	Hungary	Contest
Circle of light	Europe	Russia	Contest
IMAPP	Europe	Romania	Contest
Odessa Light Festival	Europe	Ukraine	Contest
Fête des Images	Europe	France	Mini contest
He will flee when he grows up	Europe	Italy	Mini contest
Light in Jerusalem	Middle East	Israel	Call for projects
Burj Khalifa led call	Middle East	United Arab Emirates	Call for projects
Light design contest	Middle East	Israel	Call for projects

Table 13.1. *Calls for festival projects published on the Facebook Group List of international projection mapping open call & contest 2018*

To provide an overview of the motivations of each participant and their artistic profile, a questionnaire was put online in September 2018. Twenty-one of the group members responded, allowing the following analysis. Half of the artists surveyed work alone or in pairs with sound/visual and only 10% in teams of more than six people.

Two thirds responded that they participated in between two to three contests each year. One might think that large teams participate in more contest than artists working alone, teams having more staff to handle different projects. This is not true since four of the six participants in more than five contests per year work alone or in pairs. A large majority of our artists say they have won at least one prize in a contest in the past three years. This is due to the fact that several prizes are awarded at each contest, not to mention the audience prize awarded in addition.

The reasons for participation vary. Several artists emphasize the possibility of increasing their portfolio and obtaining visibility because the videos of the screenings are relayed on an international scale. Many participate out of passion for projection mapping creation because the buildings proposed by festivals are often prestigious and have interesting architectures to work on artistically. These contests also encourage meetings between professionals in the field and create common synergies. The words *challenge, experience, artistic freedom*, and *opportunity* often appear in the answers. Finally, the need to share this artistic practice with as many people as possible was also highlighted, emphasizing the fact that projection mapping is democratic and accessible to all since the screenings take place in the public space.

More than 50% of artists finance their creative period with money earned on other projects, such as VJing, commercial projects, other activities related or not to projection mapping, or with their own funds on their free time. This explains the high number of artists working alone since they do not need to pay a team to respond to these calls for projects. Only teams with more than five people respond that they can make a living from projection mapping, including commercial projects. Most artists do not live off it or complete it with an additional activity, such as that of multimedia artist, lighting technician, university professor, visual and sound creator for the theater.

A small majority of the artists questioned prefer an artistic freedom without a theme imposed on the contests. The others explain that the theme brings, despite different approaches and styles, a global link between the different projections.

It should be noted that the panel is composed mainly of European artists, a third of Asian artists and finally a few representatives from Latin America, North America and the Middle East. This is generally representative of the origins of the participants in the contests. Indeed, Hendrik Wendler points out that he is still receiving many more proposals from artists from Germany, Austria, Spain and France as well as from a large community in Eastern European countries such as Hungary and Romania. There is a lack of

African artists on the international scene, while there is a growth in both artists and newly created festivals in Latin America.

13.4.2. *Results of the 2018 contests*

To get an idea of the growing success of these calls for contests, the 1-Min Projection Mapping explains that it has received just over 120 proposals from 39 different countries, while the Circle of Light has counted 150 from 36 nations for the 2018 edition. The Weimar festival acknowledges having had very variable results in the first years, but the number of projects received in the last editions is between 80 and 120.

Looking at the results of 15 contests held in 2018, most of which were held in Europe (Figure 13.2), we can see a strong presence of the winning European teams by including the jury and public prizes (Table 13.2 and Figure 13.3). There are five artists from Spain, four from Germany and four from Italy, three from France, 10 from Eastern Europe, Romania and Hungary mainly, and five from Asia, seven from the Americas including three from Mexico and three from Brazil; four from Russia and one from the Middle East.

It should be noted that festival audiences tend to give priority to nationals of their own country, which is true of other types of contest.

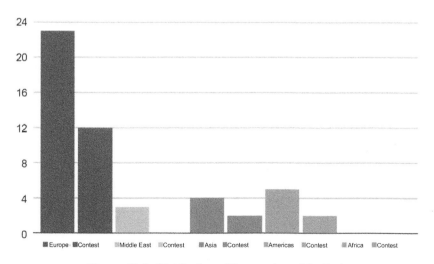

Figure 13.2. *Distribution of the number of festivals by continent in 2018*

Ranking list	Winner's country	Country where the contest is held	Festival	Artists
1	Italy	Japan	1MIN PM	Antaless visual design
2	Mexico	Japan	1MIN PM	AVA animation
3	Spain	Japan	1MIN PM	Horatu visual guerrilla
4	Hong Kong	Japan	1MIN PM	Fab asia
Public	Thailand	Japan	1MIN PM	Decide kit
1	Mexico	Germany	BERLIN festival of light A	ADhox
2	China	Germany	BERLIN festival of light A	SKGMEDIA
3	Japan	Germany	BERLIN festival of light A	NAKED INC
1	Germany	Germany	BERLIN festival of light B	MP studio
2	Hungary	Germany	BERLIN festival of light B	GLOWING BULBS
3	Romania	Germany	BERLIN festival of light B	Nomadic workshops
1	Russia	USA	Borealized	Sila Sveta
Public	USA	USA	Borealized	George Berlin
1	Russia	Russia	Circle of lights Modern	Meshsplash
2	Bulgaria	Russia	Circle of lights Modern	Elektrick me
3	Spain	Russia	Circle of lights Modern	Darklight studio
1	Germany	Russia	Circle of lights Classic	Resorb
2	Russia	Russia	Circle of lights Classic	Antimotion
3	Indonesia	Russia	Circle of lights Classic	Modar
1	Netherlands	Romania	IMAPP	Impossible vision
Public	Romania	Romania	IMAPP	Mindscape studio
1	Portugal	Ukraine	Kyiv Light festival	Mojo studio
2	Japan	Ukraine	Kyiv Light festival	Underwater

Architectural Projection Mapping Contests

3	Slovakia	Ukraine	Kyiv Light festival	Sightless
1	Spain	Spain	The crane encantada	VPM
2	Germany	Spain	The crane encantada	Videokunst
1	Spain	Spain	Luz Y Vanguardias	Video mapping pro
Public	Romania	Spain	Luz Y Vanguardias	Vali chincisan animation
1	Iran	Italy	M3D FEST	Amin Sadeghvand
2	Poland	Italy	M3D FEST	Ari Dykier
3	France	Italy	M3D FEST	Jérémy Oury
1	Italy	Italy	Novi light sensation	Giuseppe Reps
2	France	Italy	Novi light sensation	Jérémy Oury
3	Italy	Italy	Novi light sensation	Tommaso Rinaldi
1	Spain	Ukraine	Odessa Light Fest	VPM
2	Italy	Ukraine	Odessa Light Fest	Michele Puscded du
3	Brazil	Ukraine	Odessa Light Fest	John Sabbat
Public	Romania	Romania	Spotlight	Mindscape studio
1	Brazil	Brazil	SSA Mapping	Weber Bagetti
2	Brazil	Brazil	SSA Mapping	Graziella Paes & Mattheus Macedo
1	France	France	Video mapping contest	Francois Chetcuti
Public	Belgium	France	Video mapping contest	Cedric Jasmin
1	Germany	Germany	Weimar Genius Loci	Multiscalar
1	Portugal	Germany	Weimar Genius Loci	5 elements
1	Russia	Germany	Weimar Genius Loci	404.zero
1	Hungary	Hungary	Zsolnay light art	FLUID
2	Hungary	Hungary	Zsolnay light art	Eva kertesz
3	Mexico	Hungary	Zsolnay light art	Nullpixel multimedia

Table 13.2. *Festival winners published on the Facebook group List of international video mapping open call & contest 2018*

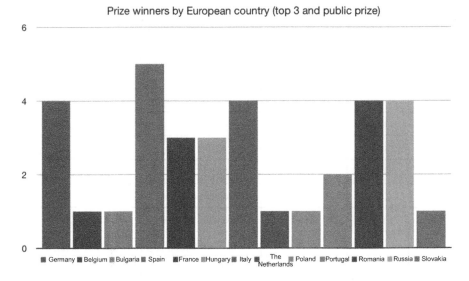

Figure 13.3. *Nationality of the European winners in the 2018 contests*

The best cash prizes that have been seen this year are those of the Luz y Vanguardias festival in Salamanca, Spain, with a first prize of €15,000, or the major IMAPP contest, where ten projects by artists chosen by the jury have been awarded €10,000 each. The festival also covered their transportation costs so that they could come and participate in the festival. Genius Loci Weimar also offered €10,000 for the three selected projects. The 1-Min Projection Mapping offers a first prize of 1,000,000 yen or about €8,000. The other festivals have price lists ranging from €500 to €5,000.

13.5. Conclusion

Spread over the year, projection mapping contests offer artistic opportunities accessible to all and promote experimentation, the discovery of new artists and the diversity of styles. However, most festivals agree that there is increasing competition with a multiplication of teams from all over the world. While Europe and Asia are home to the majority of festivals and artistic teams, the American continent is experiencing strong growth in the digital art world and American artistic proposals, particularly from South America, are multiplying. This contest is beneficial since it forces artists to innovate by proposing original concepts. Nevertheless, the increased difficulty of the contest points to the limit of this mode of participation. Indeed, to stand out, artists must spend time on their proposals and this is done, except for award winners or finalists, without financial return. Projection mapping contests are therefore limited by the contest system, which does not pay fair compensation for the work done by each competitor. Although festivals commission and directly finance artists to make unique works, if we want to move the practice of architectural projection mapping out of the spectacular event, we must urgently develop real platforms for exchange, production, distribution and creation on an international scale to make it evolve as an art in its own right.

14

Points of View: Supporting and Highlighting Projection Mapping

14.1. Video Mapping European Center according to Antoine Manier

The association Rencontres Audiovisuelles has been working on images for 20 years. Anchored in the Hauts-de-France region, it aims to provide the regional public with access to independent audio-visual content that is barely or not very visible. Born in 1998 from short film enthusiasts, it has since expanded to include various forms of digital images. The association is focused on dissemination, education, image training and support for artists. Ten years ago, Rencontres Audiovisuelles took a stronger interest in projection mapping for what this art form brings that is different. Projection mapping brings other dimensions to reality and opens up a vast field of practice for artists. The association created the Video Mapping European Center to support the development of the sector in Europe and in particular in the Hauts-de-France region. The projection mapping sector is showing itself to be unstructured in France, but also in Europe. However, it is a market with about 150 high-performance companies. In 2018, it is launching a festival dedicated to this digital art. About a third of the team is now involved in the Video Mapping Festival project.

Having appeared about 15 years ago, this projection mapping first made itself known to the public through its monumental projections on heritage, which are at the basis of many urban routes (Le Mans, Chartres, etc.), the

---

Chapter written by Marine THÉBAULT and Ludovic BURCZYKOWSKI.

enhancement of monuments (cathedrals of Amiens, Orléans, Reims, etc.) and festivities such as the Festival of Lights in Lyon. However, projection mapping has other facets than the spectacular and has strong artistic and economic potential by lending itself to various configurations, such as projections on objects or models of architecture or industry, immersive, linear, interactive mapping, etc.

The Video Mapping Festival stands out by considering projection mapping in all its forms of expression. For the 2019 edition, for example, it mixes live performance, performance and projection mapping. The team's aim is to highlight all these formats, both indoors and outdoors.

Beyond that, the name "projection mapping" is not clearly identified or recognized by the public. Compared to a similar sector such as audio-visual, projection mapping does not have a structuring framework. There are international competitions such as *1-Minute Projection Mapping* in Japan, *Art Vision* in Russia, *Mapping Contest* in Lille or *iMapp* in Bucharest, but there is no school or festival dedicated to projection mapping. However, a festival is a competitive or non-competitive program, addressed to all audiences: amateurs and professionals of the image meet and exchange. Projection mapping, which has a production line close to animation film, is also not to be confused with sound and light shows.

Projection mapping has a unifying power for audiences: 80,000 spectators attended the festival in March 2018 in Lille. It brought together families with children, contemporary art audiences, students attracted by the festive and spectacular aspect, heritage lovers, etc. The strength of projection mapping consists of its ability to reach everyone and more easily than forms of expression for which it is necessary to go to the cinema to discover them. The spectacular nature of the works and the public space facilitate access to them.

Beyond the event, the Video Mapping European Center offers training times. All year round, a team trains and supports young talents in the development of their careers by propelling them, for example, towards their first professional experience in residence.

Similarly, residences are not intended for established and identified creators. On the contrary, they support creators who already have a practice or not and who need support to move from the status of emerging practitioner to professional, but also to allow artists from similar fields to experience this artistic form (animation film artist, video game artist, etc.). The festival connects young talents and new opportunities offered by projection mapping to potential audiences and sponsors both internationally and locally. The work is ephemeral, but the festival keeps a video record of the screenings, a real business card that encourages young creators to explore new horizons. The more the festival is known, the more the programmed creators will be known. The residences welcome, for example, animation and video game professionals to go beyond the usual game of the spectacular and reflect on a new way of writing projection mapping.

According to Antoine Manier, projection mapping writing is still in its infancy and will become more structured. A grammar of projection mapping will emerge, develop and different forms of writing, linear, generative or interactive will develop. To make the analogy with cinema, the film *L'Arrivée d'un train en gare de La Ciotat*, directed by Louis Lumière in 1896, shows how much the spectacular aspect of seeing a train arrive was enough for the spectator to see a building come to life today. Projection mapping wishes to take a new look at its writing, to move towards new proposals. Storytelling could, for example, renew projection mapping. Historically, it is the worlds of VJing and communication that have taken over projection mapping. The 2018 residencies provided technical and writing support to creators in the visual and plastic arts. The idea was to give these artists from another field the keys to practice projection mapping. From now on, thinking about incorporating a cinematographic or interactive language into projection mapping could enrich the proposals in another way.

The Video Mapping European Center also offers research work carried out with the DeVisu laboratory of the Polytechnic University of Hauts-de-France. By producing reference data, the structure aims to boost research in the projection mapping sector, to better understand the evolution of the sector and to encourage creation through research.

14.2. Lighting design and sustainable projection mapping installations according to Alain Grisval

14.2.1. *Lighting designer*

Alain Grisval comes from the world of urban planning and lighting. Its subject is to ask how to create welcoming places through light. And that sell when it comes to stores. Alain Grisval wants to feel why we feel good or not in a bright space, considering the contextual nature of lighting as part of a whole. In the cities of the Nordic countries, for example, the intensity is low and uniform and you feel at home. In southern Europe, where we live at night, use is more intensive and overwhelming, but we also feel good there.

There is a link between projection mapping and the lighting designer profession. The latter has existed for about 40 years and comes from the world of entertainment, which has established a dialogue with architecture and urban planning. The lighting designers initially came mainly from the field. This is a discipline that is identified today but it was necessary to use the word or expression. This recognition took time, but the Louis Lumière schools in Lyon or the Nantes school of architecture are now training in lighting design. However, lighting designers seem cautious with projection mapping. Although in some cases very precise gobos or video projectors are used, its consideration is rather limited.

14.2.2. Durable devices

Alain Grisval thinks that we could probably surprise with more continuous projections and not only with ephemeral highlights, even if the events are a good thing. By excluding urban happening, an important adjustment must take place before an event in the case of technically heavy installations. Spontaneity is removed when, for example, public lighting is first affected by a physical operation in the field. Generating an appointment before a projection mapping event then takes away the surprise.

The transparency of the technique is a good thing, according to Alain Grisval. However, one of the problems for durable furniture in urban environments is the durability due to the bad weather, and perhaps the power of projectors to cope with the luminance of the scenes to be mapped. This has a cost and principals have problems defining with their partner a return on investment.

14.2.3. Economy

However, projection mapping is a vector for cultural, tourist and economic development. For example, the city of Chartres has opted for a lighting approach that meets a tourism challenge. It is a city that, holding a heritage, could rely on a strong attractiveness because of its proximity to Paris, from which it did not benefit in this way. Light has become one of the vectors of the attractiveness that the city has chosen through sustainable development. The night layout runs for eight months now and tour operator packages have been put in place offering a discovery tour of Northern Europe with a night in Chartres. A whole activity, catering, overnight stays, etc., did not exist at this point before this decision was taken.

14.2.4. *Legal aspect*

Legally, however, we do not only place ourselves within the limits of the laws on illuminated signs, but also within those known as "intrusive light": urban light that enters private space. It is also a question of pollution and nuisance, ecosystems for example. There are risks of glare, temperature and color issues. According to Alain Grisval, all this should be in line with the training of the actors. Alain Grisval also finds the normative approach rather uncomplicated with its generic standards. They consider "axes" or roads rather than people travelling through the city. However, the city is a system and the solutions to be found are always contextual. Alain Grisval's job, he says, is to work for people. He is for a peaceful city that respects biodiversity and believes that we can consume less while creating more meaning. Luminance, for example, is a unit that can be measured with a device, but it is very rarely mentioned and used, whereas it is fundamental: it is what the eye sees. In projection mapping, it changes every second.

14.2.5. *Identity and taste*

As a child, he was introduced to Tintin at school with a slide projector. The teacher who was playing with the projector said, "Look, Tintin is on the ceiling!" Alain Grisval is not the only one who has experienced this, he says. It seems to him that the draftsman Tronchet drew this anecdote in comics, which may have marked the beginning of mapping. Since then, he has seen maybe 70 of them, many of them in festivals. Alain Grisval believes that projection mapping is an art or artistic activity that can, with a narrative, tell a story in an urban setting and create an experience or accompany one and extend it. As an amateur of architecture and urban planning, he appreciates when projection mapping reveals, constructs and deconstructs the object on which it is projected. Just as he appreciates the dialogue with other objects in space. He prefers when the narrative is worked on. "We need narration," he says, "it's important". On the other hand, he appreciates less when the specific features of the medium are forgotten, or when the relationship is academic or too wise. He also finds that using prefabricated effects or used themes can remain pleasant, but it is necessary to use them with meaning then. It must not be visible, otherwise we will fall into the cheap effect. Time travelling though, for example, is nice but very popular. Care must be taken to ensure that this theme is used properly. What he ultimately prefers are innovative projection mappings. He likes to be surprised and also likes the subtleties that are not accessible to everyone at first reading.

14.2.6. *Interaction for all audiences*

When it comes to interacting with the public, there are cultural constraints. In France, the interaction on light is seen from the angle of overflow, and too few reflections are made on these fears. There should be a way to bring decision-makers together on this subject, so as not to reduce it to "we have never done it, so why do we have to do it?" Interactive possibilities are most likely under-exploited and undervalued according to Alain Grisval. He argues that one could very well leave the simple manipulation of a basic on/off to people's appreciation. But the problem of moderation posed by the control of various and varied users via smartphones is seen as a constraint. There are cases that have clearly shown that interaction for the general public is not interesting in its utilitarian technical aspect. But things are very different according to Alain Grisval if we look at this question in its aesthetic dimensions.

List of Authors

Julian ALVAREZ
DeVisu Laboratory
Polytechnic University of Hauts-de-France
Valenciennes
France

Hafida BOULEKBACHE
DeVisu Laboratory
Polytechnic University of Hauts-de-France
Valenciennes
France

Ludovic BURCZYKOWSKI
DeVisu Laboratory
Polytechnic University of Hauts-de-France
Valenciennes
France

Douniazed CHIBANE
DeVisu Laboratory
Polytechnic University of Hauts-de-France
Valenciennes
France

Alexandra GEORGESCU PAQUIN
TURCiT Research Group
CETT School of Tourism, Hospitality and Gastronomy
Barcelona
Spain

Sofia KOURKOULAKOU
University of Paris VIII
France

Justyna Weronika ŁABĄDŹ
University of Silesia
Katowice
Poland

Pascal LEVEL
DeVisu Laboratory
Polytechnic University of Hauts-de-France
Valenciennes
France

Nicolas LISSARRAGUE
DeVisu Laboratory
Polytechnic University of Hauts-de-France
Valenciennes
France

Jérémy OURY
Digital artist
Embrun
France

Daniel SCHMITT
DeVisu Laboratory
Polytechnic University of Hauts-de-France
Valenciennes
France

Martina STELLA
University of Paris VIII
France

Marine THÉBAULT
DeVisu Laboratory
Polytechnic University of Hauts-de-France
Valenciennes
France

Index

A

animation, 5, 20, 27, 30, 39, 42, 44, 54, 70, 90, 92, 121, 127, 135, 182, 215, 230
apparatus, 11, 20, 29, 33, 51, 52, 62, 41, 80, 115, 128
architecture, 39, 43, 46, 64, 71, 90, 123, 128, 186, 199, 216, 217, 232, 233
art, 6, 7, 11, 14–16, 18, 20, 22, 23, 26–29, 32, 33, 39, 54, 56, 63, 66, 72–79, 81, 85, 96, 102–104, 144, 149, 151, 182, 196, 199, 200, 201, 204, 208, 210, 217, 226, 227, 229, 230, 233
artistic, 18, 26, 27, 53, 56, 72, 74, 85, 96, 97, 99, 101, 103, 120, 142, 145, 149, 152, 177, 178, 181, 184, 186, 190, 194, 200, 213–217, 221, 222, 229–231, 233
artists, 4, 5, 18, 22, 28, 33, 61, 72–78, 85, 99, 102–104, 143, 147, 151, 152, 153, 194, 213–217, 219–223, 226, 227, 229, 231
automatic calibration, 107, 108, 111, 112

B, C

body, 4, 7, 14, 17, 32, 40, 41, 43, 47, 80, 86, 97, 156, 171
cinema, 5, 9, 11, 12, 15, 30–34, 38–40, 42–44, 47, 48, 51, 52, 56, 58, 59, 61, 73, 102, 142, 144, 150–153, 231
communication, 54, 123, 163, 164, 172, 175, 178–180, 182, 231
contest, 213–223, 226
creation, 39, 58, 73, 85, 120, 127, 141, 153, 177, 213, 231
creative
 dynamic, 85
 process, 86, 88, 91, 95–97, 99, 100–103, 142, 146, 148, 188, 195
cultural heritage, 178–183, 186, 194, 195, 199–201, 211

D, E

dimension, 90, 131, 204
diversity, 217
environment, 8, 43, 55–57, 60, 64, 65, 66, 72, 76, 80, 81, 85, 86, 88, 89, 95, 100–102, 104, 117, 118, 121, 137, 138, 153, 154, 157, 167, 170

experience, 38, 44, 45, 48, 60, 62, 80, 85–87, 91, 96, 100, 103, 121, 144, 145, 150, 154–156, 178, 180, 181, 186, 187, 193, 194, 203, 208, 213, 214, 222, 226, 233
experiment, 8, 17, 69, 70, 73, 97, 144, 147

F

façade, 4, 6, 42, 46, 47, 73, 74, 118, 123–125, 128, 151, 153, 154, 181, 184, 186, 200, 206, 208–210, 214, 215
factory of the future, 161–164, 169, 171, 172, 175
festival, 213–219, 222, 223, 226, 227

H, I

heritage, 63, 155, 177
illusion, 8–12, 17, 26, 30, 38, 39, 41, 42, 44, 47, 104, 119, 146
immersion, 8, 11, 14, 38, 40, 47, 71, 77, 80, 143, 194
installation, 33, 40, 42–45, 53, 54, 56, 57, 61–63, 70, 71, 73, 75, 79, 81, 120, 124, 154, 182, 186, 191, 192, 231
interactivity, 153
international, 214, 222, 223, 227

M

mediatecture, 181
mediation, 58, 63, 102, 154, 177, 180–182, 184–186, 188, 189, 191, 194–196, 199, 202–204, 210, 211
medium, 22, 42, 43, 74, 77, 78, 102, 177, 196
memory, 14, 15, 34, 63, 65, 78, 86, 178, 193, 195, 211
model, 90, 98, 128–130, 134, 137, 164, 165, 170, 171, 213, 215

museum, 52, 53, 58, 76, 81, 79, 97, 120, 121, 154, 178, 182–184, 188–194, 201, 204, 214, 217
music, 28, 79, 141–143, 145–147, 184, 199, 204, 215

O, P

organizers, 214, 216, 217
perception, 10, 11, 14, 22, 38, 39, 41, 43–48, 52, 53, 62, 65, 144, 153, 156, 165, 210
perspective, 11, 12, 15, 22–24, 26, 27, 40, 41, 51, 52, 59, 61, 62, 128
photogrammetry, 127–131, 134, 135, 137–139
practice, 18, 54, 55, 64, 66, 74–77, 79, 81, 85, 88, 89, 91, 93, 103, 149, 177, 214, 222, 227, 230

R

reality
   augmented, 11–13, 38, 39, 44, 47, 48, 76, 80, 122, 161, 165–167, 169–172, 175, 179, 189, 203
   mixed, 138, 167
   virtual, 38, 39, 45, 62, 76, 80, 117, 121, 166
reconstruction, 127–130, 132–134, 137, 138
residence, 85–88, 89, 91, 92, 94–96, 100–104, 230, 231

S, T

software, 56, 58, 75, 85, 91, 93–95, 98, 100, 101, 128, 129, 131, 134, 137, 161, 164, 205
sound, 20, 28, 70, 80, 141–149, 151, 154, 195, 215, 216, 222
spectacle, 5, 6, 26, 28–31, 39, 40, 46, 72, 75, 77, 104, 119, 143, 150, 151, 153, 178, 182, 191, 195, 204, 215, 230, 232

spectator, 12, 17, 30, 31, 38–48, 57, 59, 61–63, 65, 70, 80, 90, 141, 143–145, 149, 153–156, 187, 206, 216, 230, 231
sponsors, 216, 231
surface, 4, 7, 12, 22, 25, 43, 51, 54–63, 65, 79, 90, 118, 119, 123, 137, 142, 146, 153, 156, 170, 172, 174, 178, 181, 210, 216
symbolic, 13, 14, 16, 41, 51, 52, 61, 65, 66, 177–181, 184, 185, 194, 211
technologies, 32, 38, 39, 79, 80, 116, 118, 122, 125, 143, 156, 162, 169–172, 193
theater, 15, 30, 41, 206

V, W

video
  game, 115, 156, 231
  projection, 117–123, 125, 126
  projector, 73–77, 85, 117, 118, 126
writing, 26, 28, 40, 45, 47–49, 63, 65, 72, 73, 79, 80, 94, 98, 99, 117, 120, 121, 125, 130, 138, 142, 144–147, 154, 156, 164, 169, 171, 186, 187, 218, 234

Other titles from

in

Science, Society and New Technologies

2019

BRIANÇON Muriel
The Meaning of Otherness in Education: Stakes, Forms, Process, Thoughts and Transfers
(Education Set – Volume 3)

DESCHAMPS Jacqueline
Mediation: A Concept for Information and Communication Sciences
(Concepts to Conceive 21st Century Society Set – Volume 1)

DOUSSET Laurent, PARK Sejin, GUILLE-ESCURET Georges
Kinship, Ecology and History: Renewal of Conjunctures
(Interdisciplinarity between Biological Sciences and Social Sciences Set – Volume 3)

DUPONT Olivier
Power
(Concepts to Conceive 21st Century Society Set – Volume 2)

FERRARATO Coline
Prospective Philosophy of Software: A Simondonian Study

GUAAYBESS Tourya
The Media in Arab Countries: From Development Theories to Cooperation Policies

HAGÈGE Hélène
Education for Responsibility
(Education Set – Volume 4)

LARDELLIER Pascal
The Ritual Institution of Society
(Traces Set – Volume 2)

LARROCHE Valérie
The Dispositif
(Concepts to Conceive 21st Century Society Set – Volume 3)

LATERRASSE Jean
Transport and Town Planning: The City in Search of Sustainable Development

LENOIR Virgil Cristian
Ethically Structured Processes
(Innovation and Responsibility Set – Volume 4)

LOPEZ Fanny, PELLEGRINO Margot, COUTARD Olivier
Local Energy Autonomy: Spaces, Scales, Politics
(Urban Engineering Set – Volume 1)

MARTI Caroline
Cultural Mediations of Brands: Unadvertization and Quest for Authority
(Communication Approaches to Commercial Mediation SET – Volume 1)

METZGER Jean-Paul
Discourse: A Concept for Information and Communication Sciences
(Concepts to Conceive 21st Century Society Set – Volume 4)

MICHA Irini, VAIOU Dina
Alternative Takes to the City
(Engineering, Energy and Architecture Set – Volume 5)

PÉLISSIER Chrysta
Learner Support in Online Learning Environments

PIETTE Albert
Theoretical Anthropology or How to Observe a Human Being
(Research, Innovative Theories and Methods in SSH Set – Volume 1)

PIRIOU Jérôme
The Tourist Region: A Co-Construction of Tourism Stakeholders
(Tourism and Mobility Systems Set – Volume 1)

PUMAIN Denise
Geographical Modeling: Cities and Territories
(Modeling Methodologies in Social Sciences Set – Volume 2)

WALDECK Roger
Methods and Interdisciplinarity
(Modeling Methodologies in Social Sciences Set – Volume 1)

2018

BARTHES Angela, CHAMPOLLION Pierre, ALPE Yves
Evolutions of the Complex Relationship Between Education and Territories
(Education Set – Volume 1)

BÉRANGER Jérôme
The Algorithmic Code of Ethics: Ethics at the Bedside of the Digital Revolution
(Technological Prospects and Social Applications Set – Volume 2)

DUGUÉ Bernard
Time, Emergences and Communications
(Engineering, Energy and Architecture Set – Volume 4)

GEORGANTOPOULOU Christina G., GEORGANTOPOULOS George A.
Fluid Mechanics in Channel, Pipe and Aerodynamic Design Geometries 1
(Engineering, Energy and Architecture Set – Volume 2)

GEORGANTOPOULOU Christina G., GEORGANTOPOULOS George A.
Fluid Mechanics in Channel, Pipe and Aerodynamic Design Geometries 2
(Engineering, Energy and Architecture Set – Volume 3)

GUILLE-ESCURET Georges
Social Structures and Natural Systems: Is a Scientific Assemblage Workable?
(Social Interdisciplinarity Set – Volume 2)

LARINI Michel, BARTHES Angela
Quantitative and Statistical Data in Education: From Data Collection to Data Processing
(Education Set – Volume 2)

LELEU-MERVIEL Sylvie
Informational Tracking
(Traces Set – Volume 1)

SALGUES Bruno
Society 5.0: Industry of the Future, Technologies, Methods and Tools
(Technological Prospects and Social Applications Set – Volume 1)

TRESTINI Marc
Modeling of Next Generation Digital Learning Environments: Complex Systems Theory

2017

ANICHINI Giulia, CARRARO Flavia, GESLIN Philippe,
GUILLE-ESCURET Georges
Technicity vs Scientificity – Complementarities and Rivalries
(Interdisciplinarity between Biological Sciences and Social Sciences Set – Volume 2)

DUGUÉ Bernard
Information and the World Stage – From Philosophy to Science, the World of Forms and Communications
(Engineering, Energy and Architecture Set – Volume 1)

GESLIN Philippe
Inside Anthropotechnology – User and Culture Centered Experience
(Social Interdisciplinarity Set – Volume 1)

GORIA Stéphane
Methods and Tools for Creative Competitive Intelligence

KEMBELLEC Gérald, BROUDOUS EVELYNE
Reading and Writing Knowledge in Scientific Communities: Digital Humanities and Knowledge Construction

MAESSCHALCK Marc
*Reflexive Governance for Research and Innovative Knowledge
(Responsible Research and Innovation Set - Volume 6)*

PARK Sejin, GUILLE-ESCURET Georges
*Sociobiology vs Socioecology: Consequences of an Unraveling Debate
(Interdisciplinarity between Biological Sciences and Social Sciences Set – Volume 1)*

PELLÉ Sophie
*Business, Innovation and Responsibility
(Responsible Research and Innovation Set – Volume 7)*

2016

BRONNER Gérald
Belief and Misbelief Asymmetry on the Internet

EL FALLAH SEGHROUCHNI Amal, ISHIKAWA Fuyuki, HÉRAULT Laurent, TOKUDA Hideyuki
Enablers for Smart Cities

GIANNI Robert
*Responsibility and Freedom
(Responsible Research and Innovation Set – Volume 2)*

GRUNWALD Armin
*The Hermeneutic Side of Responsible Research and Innovation
(Responsible Research and Innovation Set – Volume 5)*

LAGRAÑA Fernando
E-mail and Behavioral Changes: Uses and Misuses of Electronic Communications

LENOIR Virgil Cristian
Ethical Efficiency: Responsibility and Contingency
(Responsible Research and Innovation Set – Volume 1)

MAESSCHALCK Marc
Reflexive Governance for Research and Innovative Knowledge
(Responsible Research and Innovation Set – Volume 6)

PELLÉ Sophie, REBER Bernard
From Ethical Review to Responsible Research and Innovation
(Responsible Research and Innovation Set – Volume 3)

REBER Bernard
Precautionary Principle, Pluralism and Deliberation: Sciences and Ethics
(Responsible Research and Innovation Set – Volume 4)

VENTRE Daniel
Information Warfare – 2nd edition

Printed and bound by CPI Group (UK) Ltd, Croydon, CR0 4YY
04/09/2023